Charles Warrenne Allen, Jacob Sobel

Handy book of medical progress

A lexicon of the recent advances in medical science

Charles Warrenne Allen, Jacob Sobel

Handy book of medical progress
A lexicon of the recent advances in medical science

ISBN/EAN: 9783337223366

Printed in Europe, USA, Canada, Australia, Japan

Cover: Foto ©berggeist007 / pixelio.de

More available books at **www.hansebooks.com**

HANDY BOOK

OF

MEDICAL PROGRESS

A LEXICON OF THE RECENT ADVANCES
IN MEDICAL SCIENCE

BY

CHARLES WARRENNE ALLEN, M.D.

CONSULTING DERMATOLOGIST TO THE RANDALL'S ISLAND HOSPITALS ; CON-
SULTING SURGEON (GENITO-URINARY) TO THE CITY HOSPITAL ;
ATTENDING SURGEON (DEPARTMENT OF SKIN) GOOD
SAMARITAN DISPENSARY, ETC.

AND

JACOB SOBEL, A.B., M.D.

DERMATOLOGICAL ASSISTANT AT THE GOOD SAMARITAN DISPENSARY ; MEMBER
OF THE MEDICAL SOCIETY OF THE COUNTY OF NEW YORK,
HARLEM MEDICAL ASSOCIATION, ETC.

NEW YORK
WILLIAM WOOD AND COMPANY
MDCCCXCIX

PREFACE.

So rapidly do the advances in the various branches of medical science multiply, that one finds it an almost impossible task to keep abreast of the times without the aid of what might be called concentrated literature. In current publications many terms and neologisms are employed long before they find their way into text-books, encyclopædias, and dictionaries. For these reasons the authors have been led to compile an alphabetically arranged volume, giving so far as possible the meaning of new terms and value of the new discoveries in the realms of practical medicine. The book includes the more recent and novel names of diseases, tests, methods, drugs, therapeutic and surgical suggestions, etc.

It would have been impossible and unwise in a work of this kind to attempt even an enumeration of all the remedies recently introduced. In other branches as well, omissions have occurred, since to incorporate all the novelties of medicine, even for a single year, would call for a volume whose size would defeat the ends of condensation and portability. While the authors cannot individualize, they desire to express in a general way their indebtedness to the numerous foreign and domestic journals, periodicals, and monographs from which they have freely drawn. When expedient, the source of information has been acknowledged, with the name of the author, journal, and date. When a given term is not in itself strictly new, there will usually be found something of a progressive nature in the subject-matter placed under the title.

If this little work should furnish a means of conveniently and rapidly acquiring information relative to the newer advances, the endeavors of the authors will have been fulfilled.

NEW YORK, May 1, 1899.

HANDY BOOK

OF

MEDICAL PROGRESS.

Abasia.—A term applied by Blocq to motor incoördination in walking. Observed most frequently in hysteria. Muscular strength and sensation are not disturbed.

Abulia.—An absence or defect of will-power, observed in conditions of hysteria, neurasthenia, melancholia, and morphinism.

Acanthosis Nigricans, or Keratosis Nigricans (Kaposi).—Grayish-brown or brownish-black warty growths occurring on the back of the neck and in the pectoral and axillary folds. It is a disease of adult life, and the patients as a rule die within a comparatively short time after its development, usually of carcinoma of some of the abdominal viscera, especially the stomach.

Acetal.—Hypnotic in doses of ℨ ss. to ℨ ii. It contains one molecule of aldehyde and two of alcohol.

Acetanilid (Antifebrin).—Analgesic, antirheumatic, and antithermic. The product of acetic acid on aniline. A pure white crystalline powder, of silky lustre, scarcely soluble in cold water, giving a slight burning taste to the tongue. May be given in wafers, in elixir, or in strong wine. Dose, gr. ii. to viii.

1

It has also been used as an antiseptic powder, ointment, and as a gauze, for ulcers, wounds, etc.

Caution.—This drug may produce cyanosis and sudden collapse.

Actol.—See Silver Lactate.

Acetonuria.—A pathological condition, in which the urine shows the presence of acetone. Von Jaksch classifies the following forms of pathological acetonuria:

1. The febrile. 2. The diabetic. 3. Acetonuria in certain forms of carcinoma which have as yet not led to inanition. 4. The acetonuria of inanition. 5. The occurrence of acetonuria in psychoses. 6. Auto-intoxication. 7. Digestive disturbances. 8. Chloroform narcosis.

According to Knapp, death of the fœtus is accompanied by the appearance of acetone in the mother's urine.

Tests.—To a few cubic centimetres of urine add a few drops of a freshly prepared concentrated solution of nitroprusside of sodium, and also a few drops of strong sodium or potassium hydrate. The fluid takes on a red color, which rapidly becomes paler, but assumes a purple-red or violet-red color upon the addition of acetic acid. If there be no acetone present, the fluid does not become purple-red upon the addition of acetic acid (Legal).

To a few cubic centimetres of urine add a few drops of potassium hydrate and Lugol's solution. If acetone be present, a precipitate of iodoform crystals, with a distinct iodoform odor, forms in a few minutes (Lieben).

See also Chautard's Test.

Acetphenetidin.—Synonymous with phenacetin and the preferable term for prescription writing.

Achillodynia (Achillobursitis).—A condition which consists in the inability to walk or stand erect on account of the existence of pains situated at the insertion of the tendo Achillis. The pain is generally due to the inflammation of a bursa situated between the tendon and the tuberosity of the os calcis.

Achroacytosis.—The replacement of glandular tissue by lymph cells.

Achromatopsia.—Partial or complete color-blindness. Of occasional occurrence in hysteria.

Achylia Gastrica.—Also called atrophy of the stomach, anadenia ventriculi, and phthisis ventriculi. This name was first used by Dr. Einhorn for a condition of the stomach in which there was destruction of the glandular structures with resulting secretory disturbances, absence of free and combined hydrochloric acid, pepsin, and rennet. The motility of the stomach is perfectly intact, thus distinguishing it from carcinoma. As a result of the condition there is no stagnation, fermentation, or lactic-acid formation, and the patient may enjoy comparatively good health for a number of years, the greater part of digestion being done by the intestinal juices.

Acrodynia.—Same as pellagra.

Acromegaly, or Acromegalia. — A disease first described by Marie and characterized by a considerable enlargement of the hands (spade hand) and feet, a thickened and plump condition of the face with enlargement of the nose, lips, tongue, superior and inferior maxilla. The lower jaw may become so much enlarged as to project beyond the upper (prognathous). It is a disease of adult life and seems to occur more frequently in women, in whom the first symptom may be a cessation of menstruation. The patients often complain of severe pains in the head and extremities. The spinal column may show kyphosis late in the disease; sugar may be detected in the urine, and diminution in vision (optic neuritis) and hemianopsia are common enough. The disease begins as a rule in the hands, is insidious in onset and slow and progressive in its course, death usually being due to intercurrent disease or cachexia.

The pathology of the disease is not very well understood. Most cases seem to depend upon an enlargement of the hypophysis or pituitary body; in other instances enlargement of the thymus and thyroid has been found.

Acroparæsthesia.—A name first used by Fr. Schultze for uncomfortable and painful sensations at the tips of the extremities. It is a disease of adult life and is observed more frequently in women than in men. The sensory disturbances consist in pricking, burning, sticking, and tearing, and are situated principally in the fingers and fingertips. The feet and toes are less frequently attacked. Both hands are as a rule affected, occasionally one is more involved than the other. The motor power, reflexes, and sensibility are intact. Occasionally vaso-motor disturbances—cyanosis, coldness, hyperidrosis—are observed.

Acrophobia.—Fear of high places.

Addison's Disease.—An affection due to a lesion of the suprarenal capsules or abdominal sympathetic, and characterized by a profound asthenia, cardiac weakness, debility, emaciation, gastric irritability, and a bronze coloration of the skin and mucous membranes. The pigmentation is deeper on the exposed parts—face, hands, neck—and in those regions in which pigmentation is more intense under normal condition, viz., the scrotum and breasts. The mucous membranes most frequently involved are the lips, cheeks, tongue, conjunctiva, and vagina. The prognosis is bad, death usually occurring in from a few weeks in acute cases to a few years in the more prolonged ones.

Although usually a disease of adult life, Dezerot (*Klin. therap. Wochen.*, September 18, 1898) has observed three cases in children, and has collected from literature forty-eight instances occurring in children from seven days to fourteen and a half years old.

Adenoid Face.—A stupid and semi-idiotic expression with a long, high nose and flattening of the bridge, narrow nostrils, drooping jaw, open mouth, irregularity of the upper teeth, and broadening between the eyes.

Siebenmann claims that the narrow, high, arching palate has no connection with the existence of adenoids.

Adonidin.—A glucoside of Adonis vernalis, used in regulating the heart in doses of gr. $\frac{1}{6}$ to $\frac{1}{3}$ in pill-form.

Adrenals.—Synonymous with the suprarenal capsules.

Aeroductor.—An invention of Wiedmann to relieve asphyxia in the foetus when the aftercoming head is retained.

Agar-Agar.—Recommended by Gallois for inflammatory skin affections, especially erysipelas. He uses a one-per-cent watery solution to which is added 0.1 per cent perchloride of mercury and tartaric acid. It forms a thin coating, which dries rapidly. It is inexpensive.

Agaricin.—The active principle from agaric. It consists of a white crystalline powder and is used in pill-form for the night sweats of phthisis. The dose is gr. $\frac{1}{12}$ to $\frac{1}{8}$. It also checks the bronchial secretion and dries up the milk.

Agathin (Salicyl - methyl - phenylhydrazone). — Antirheumatic agent. Dose, gr. v. to gr. x. t.i.d.

Ageusia.—Absence or loss of the sense of taste.

Agoraphobia.—Dread or fear of open places and spaces.

Agraphia.—There is a sensory and a motor agraphia, the former being the inability to express an idea in writing, the latter consisting in the loss of voluntary writing.

Aiodine.—A new preparation of thyroid obtained by Schoerges (*Nouveaux Remèdes*, August 24, 1898). It results from precipitating with tannin the iodo-albuminates, bases, and mucous substances of the thyroid gland. It is said to keep better than thyroid.

Lanz (*Berliner klinische Wochenschrift*, April 25, 1898) has found it useful after thyroidectomy and in hyperplastic goitre.

Airoform.—Another name for airol.

Airol (Oxyiodogallate of Bismuth).—One of the many substitutes for iodoform. A dark-green powder, odorless, non-poisonous, non-irritating, having antiseptic, analgesic, and styptic properties, increasing granulations, and diminishing secretions.

Aktinography (same as skiagraphy).—A term used especially in Germany.

Albolene.—A liquid petroleum product used in nose and throat affections. It serves as an excellent vehicle for menthol, cocaine, benzoic and boric acids, gaultheria, etc. It is generally used with a special atomizer.

Albumin: Tests for.—(*a*) Truax. Pour three to five cubic centimetres of urine into a test-tube and gently add half the amount of alcohol. If albumin be present, a white line will form between the two fluids, similar to the albumin ring of Heller's test (see page 81). Not reliable.

(*b*) Trichlor-acetic acid. Take two small test-tubes, each having a foot. Into one place 5 c.c. of the urine as voided and put it aside; into the other place the same amount after it has been filtered. Into the latter drop slowly the reagent, drop after drop, until opalescence or flocculency is noticed. If after the addition of thirty drops (2 c.c.) of the reagent no opalescence or sedimentation occurs, it may be concluded that the specimen of urine is free from albuminous substances. Finally, compare the urines in the two test-tubes with each other, and note carefully if the turbidity in the urine which was subjected to the reagent is of a greater degree than that of the unfiltered urine. An increased turbidity is invariably indicative of the presence of some proteid material. The trichlor-acetic acid test reacts with mucin and other albuminous elements, and is trustworthy only in detecting albumin as a class, and not serum albumin *per se.*—*Heinrich Stern.*

(*c*) "Sulpho test" of Stein. Pour a small quantity of

filtered urine into a test-tube and add a few crystals of sulpho-salicylic acid. Shake the tube, and if albumin be present, a white homogeneous precipitate will form immediately.

See also magnesium-nitric test.

Alcarnose.—A mixture of albumose and maltose. A basis for nutritive enemata. It has no smell and a faintly sweet taste.

Alcohol as an Antidote for External Carbolic-Acid Poisoning.—This substance was first recommended and used for this purpose by Dr. Seneca D. Powell, of New York. It is said to act as a thorough antidote, preventing pain, blistering, and a deep escharotic effect. It is also used as an internal antidote in the form of whiskey, brandy, etc. Zangger has recently advocated alcohol lotions or baths for burns, furuncles, felons, etc.

Aleppo Boil.—A communicable disease observed in the natives of Aleppo, beginning in the face and uncovered parts of the body as a large acne nodule which increases to the size of a twenty-centime piece. The nodule is not painful nor reddened, and becomes covered with a crust, beneath which an ulcerated surface appears at the third or fourth month. This is followed by cicatrization with a resulting permanent depressed scar. It is generally single, though sometimes there are eight to ten of them. Complications do not arise. It has been variously attributed to micrococci, staphylococci, and streptococci. Treatment is soothing and protective.

Aleuronat.—An albuminous substance to whose value

Ebstein has drawn attention as a food in diabetes. It is prepared by a special process from wheat. It is a light yellow-brown powder containing eighty to ninety per cent of albumin and seven per cent of carbohydrates. It may be obtained from R. Hundhausen, of Hamm, in Westphalia. It is a cheap substitute for ordinary flour.

Algometer.—An instrument for the measurement or comparative estimation of pain. One method is by an induced current of electricity. A temporal algometer has been devised by A. MacDonald (*Psychological Review*, 1898). It is pressed against the temple until pain is elicited, the amount of pressure being registered on a scale.

Alkaptonuria.—Alkaptone is a resinous substance of light brown color which appears in the urine under a variety of conditions and seems to be without any particular pathological significance. The characteristic of such a urine is that it is colorless when evacuated, but becomes very dark, even black, upon exposure to the air. The greatest interest of alkaptone lies in the fact that it may be mistaken for glucose, since it reduces Fehling's solution. It does not, however, ferment like glucose, nor does it respond to the more delicate phenylhydrazin and polariscopic tests, and on adding an alkali and shaking, a reddish-brown color appears almost instantaneously. The condition is one of importance in life-insurance examinations.

Allocheiria.—Inability to localize the seat of pain. Thus, if pricked with a pin on one foot, the patient may feel it in the other. This phenomenon is observed in tabes.

Alpha-Eigon.—An iodine compound of albumin (twenty per cent).

Alphasol.—This is a proprietary preparation, used as an antiseptic for gargles, nasal douche, sprays, etc.

Alumnol (Aluminium Naphthol-Sulphate).—A soluble salt of aluminium and naphthol-sulphonic acid, containing about fifteen per cent of sulphur and five per cent of aluminium. It is a non-hygroscopic, grayish-white powder, and has been used in gonorrhœa (one- to two-per-cent. solution) and in various combinations for a variety of skin diseases, being antiseptic and astringent and said to possess the power of penetrating the tissues.

Amaurotic Family Idiocy (Sachs).—A family disease, observed most frequently in children of the Jewish faith, and characterized by a psychical defect, amounting at times to total idiocy, which is observed in the early months of life, weakness of all the extremities, even to total paralysis of a spastic or flaccid nature, diminution of vision, or total blindness, and normal, increased, or diminished reflexes. The fundus of the eye shows in the macula lutea region a cherry-red spot with a surrounding white halo. Most children die of marasmus before the second year of life. The first symptoms appear between the third and seventh month. Its hereditary character while probable has as yet not been definitely settled.

Ammonol.—This is supposed to be a coal-tar derivative. According to the chemical examination of George M. Beringer (*American Journal of Pharmacy*, vol. lxix., page 150), it is composed of

Acetanelid	10 gm.
Sodium bicarbonate	5 "
Ammonium bicarbonate	5 "
Metanil yellow	0.005 "

It is a powder of a faint yellow color, with a strong ammoniacal odor and of slightly crystalline composition. It is said to be an analgesic, antipyretic, and stimulant, and its composition certainly points in these directions. It has been used in migraine, dysmenorrhœa, gastralgia, and dyspeptic disorders. It may be combined with salicylic acid, lithia, bromides, etc. The dose is gr. v. to xv.

Amusia.—Tone deafness. A condition similar to that of aphasia, in which the power of distinguishing tunes and tones is lost. Thought by Knauer to be a brain intoxication, due to functional derangement of the thymus.

Amyloform.—A patent preparation formed by combining formaldehyde with starch and used as a substitute for iodoform. It consists of a white odorless and tasteless powder and is non-toxic. It has been used in suppurating wounds, osteomyelitis, and varicose ulcers with good results. It is an antiseptic and deodorizer, and is inexpensive.

Amyloiodoform.—A combination of amyloform and iodine. Used as a substitute for iodoform.

Anabasis.—The beginning of a disease.

Anæsthesia, General.—Schleich (Berlin) has shown that the higher the boiling-point the smaller the amount of anæsthetic required.

The following combinations may be prepared:

	Mixt. I.	Mixt. II.	Mixt. III.
Chloroform................	45	45	30
Petroleum ether............	15	15	15
Sulphuric ether.............	180	150	80
Boiling point....	38° C.	40° C.	42° C.

In operations of twenty minutes' duration use Mixture I.; in more prolonged operations, Mixtures II. or III.

Advantages claimed: smaller amount required, immediate awakening after operation, less after-effects.

See Schleich method.

Analeptol (Analepsis = recovery or convalescence).—A proprietary name given to a tonic preparation, each drachm of which, according to the manufacturers, represents phosphorus $\frac{1}{100}$ gr., nux vomica extract $\frac{1}{6}$ gr., cinchona 2 gr., coca leaves 1 gr., combined with aromatics. Recommended by the promoters in convalescence, physical or mental exhaustion, etc.

Analgen.—An analgesic and anti-neuralgic which is said to color the urine red. Dose, gr. vii. to xv.

Analgesin.—Another name used in France for antipyrin.

Anarthria.—A disturbance of speech due to the involvement of the muscles of articulation such as is seen in bulbar or glosso-labio-laryngeal paralysis.

Anasin.—Said to be an aqueous solution of tri-chlor-pseudo-butyl-alcohol or aceto-chloroform. Used as an hypnotic and anæsthetic.

As an hypnotic the dose is gr. vii. to xv. A one-per-cent. solution is said to possess the same anæsthetic properties as a two-per-cent cocaine solution. It has been used in the eye, larynx, pharynx, and nasal mucous membrane.

Anastasis.—The period of convalescence.

Angina Ludovici (Ludwig's Angina, or Infectious Submaxillary Angina).—An acute disease which involves the areolar tissue surrounding the submaxillary gland, and gives rise to marked induration and suppuration of the tissues, with grave constitutional disturbances. The disease begins with local discomfort which lasts three or four days, and the neck becomes painful and swollen. The patient experiences difficulty in speech and opening the mouth, the tongue becomes swollen, there is great salivation and occasionally dysphagia or dyspnœa. The temperature may be high, and, unless operative intervention is instituted, death may result with septic symptoms. It is an unusual but severe form of cellulitis of the neck.

Anilipyrin (Acetanilid one part, Antipyrin two parts). —Dose, gr. v. to xv. Used in influenza and rheumatism.

Animal Extracts.—Since thyroid extract was first advocated in the treatment of myxœdema a great variety of similar preparations have sprung up, notably suprarenal, pituitary, thymus, splenic, hepatic, testicular, ovarian, spermatic, as well as cerebrin, medullin, and others. Reports more or less favorable continue to be published regarding this novel method of medication. Each extract will be treated under its appropriate heading.

Anthrarobin.—A compound procured by Liebermann and very similar to chrysarobin in both appearance and therapeutic indications. It is used in the form of powder, ointment, or in collodion and traumaticin, in strength of from five to ten per cent, for psoriasis, seborrhœal eczema, tinea circinata, and tinea versicolor.

Antikamnia.—A white, odorless, tasteless powder, insoluble in water, but soluble in alcohol and ether. It acts, as an antipyretic and analgesic in doses of from three to ten grains. It may be given in powder or tablet form, combined with salol, codeine, phenacetin, etc.

It is a secret American proprietary remedy, supposed to contain, according to a published formula (*Pharm. Rundschau*)—

℞ Acetanilid.....	47–86 parts.
Sodium bicarbonate	14–50 "
Tartaric acid........................	3–6 "
Caffeine.....................................	3–10 "

Antinervin (Salicyl-bromanilinid).—A combination of ammonium bromide and salicylic acid, each one part, and acetanilid two parts. Employed as a sedative in angina pectoris, articular rheumatism, and typhoid fever.

Dose, gr. iii. to v. every two to three hours in capsule, or gr. x. may be given as a single dose.

Antinosin, or Tetra - iodo - phenol - phthalein. — The sodium salt of nosophen—a blue powder, soluble in water, similar in action to nosophen, but may be used in solution, especially in nose, ear, and throat work. It is non-irritating and non-toxic.

Advocated in one- to three-per-cent solution for bladder irrigation.

Antiscabin.—A mixture of balsam of Peru, soap, glycerin, beta-naphthol, boric acid, and alcohol. Used in scabies.

Antisepsin.—A bromine substitution compound of acetanilid. Antineuralgic and antirheumatic. Dose, gr. v. to x. daily.

Antiseptin.—

R Zinci sulphatis 85.0
Acidi boracici............................... 10.0
Zinci iodidi,
Thymol....................................āā 2.5

Antiseptol (Cinchonine Iodo-sulphate).—An odorous brown powder, recommended as a substitute for iodoform. It contains half its weight of iodine, and is insoluble in water but soluble in alcohol and chloroform.

Antistreptococcic Serum (Marmorek).—This serum is said to be an antitoxin for all *true* streptococcic infections, that is, for those cases of toxæmia in which streptococci are found in the blood. It has been used in erysipelas, phlegmon, septic wounds, and puerperal septicæmia with only occasional benefit. It has been recommended for cases of mixed infection such as are seen in scarlet fever, diphtheria, and tuberculosis. It is put up in vials of 10 c.c., and may be procured from the New York Board of Health or the Pasteur laboratory. It may be used per rectum, O'Connor reporting gratifying results in scarlet fever.

Antistaphylococcus Serum.—This has been used by Moritz in malignant endocarditis with success.

Antitoxins.—These are substances which have a tendency to counteract the effects which the toxins of various micro-organisms exert upon the body. A typical example of this class is *diphtheria antitoxin*. This is produced by injecting, in gradually increasing doses, into the body of a healthy horse the by-product of the diphtheria bacilli, that is, the toxins. The horse at first reacts with fever and general indisposition, but as the dose is gradually increased he

becomes immune as it were, and does not react to the
largest of doses. The serum of his blood is then used as
antitoxin, after going through various processes of sterilization and filtration. This antitoxin, in order to produce
its most desired effects, must be used early in the disease,
preferably during the first twenty-four hours. The tendency of the present day is to use concentrated serums, that
is to say, those which represent the greatest number of units
in the smallest amount of fluid. From 1,000 to 8,000 units
in all may be injected according to the age of the child, the
severity and stage of the disease, and the existence or absence of stenotic symptoms. The favorite sites for injection are the interscapular region, the abdomen, chest, and
outer side of thigh; the procedure should be slowly done
and the fluid be injected well into the tissues. It is not advisable to massage the region after injection. If there be
no improvement in from eight to twelve hours, the initial
dose is to be repeated as often as necessary. For a working
basis it may be said, give: 1,000 to 1,200 units in young
children; 1,200 to 1,500 units in older children; 2,000 to
3,000 units in croup cases.

The site of injection, the hands of the physician, and
the syringe must be scrupulously clean. Any syringe will
answer the purpose, the simpler the better. One made
entirely of glass is to be preferred on account of the ease
with which it can be sterilized. Antitoxin has been administered per os and per rectum, but not with as good results
as hypodermatically. J. E. O'Connor gives the following
advantages of rectal over hypodermatic serotherapy: (a)
No prejudicial interference ; (b) solid serums can be used;
(c) it is simple and expeditious and the absorbed serum is

purer. Apart from its curative and specific effect, anti-toxin is used for immunizing purposes in doses of 200 to 300 units. This period of immunity lasts for from three to four weeks. Antitoxin rashes are occasionally observed after its use, and assume a scarlatiniform, urticarial, ery-thematous, or morbilliform type.

In New York, the Board of Health supplies physicians with antitoxin without charge, for use among the poor. Antitoxin of reliable quality may be obtained from the New York Board of Health, Parke, Davis & Co., Mul-ford & Co., and the imported serums of Behring and Aronson may also be procured at the shops. Extra po-tent serum in dry form can now be obtained.

Besides the diphtheria antitoxin there are antitoxins of tuberculosis, tetanus, hydrophobia, streptococcus and staphylococcus infection, erysipelas, snake bites, yellow fever, plague, etc. Antitoxins for pneumonia and typhoid fever are now being experimented with.

Antipyrin.—Introduced by Knorr. A white crystalline powder, very soluble in water, less so in alcohol or gly-cerin. It is analgesic, antipyretic, antirheumatic, and hæmostatic in its action. The dose is from gr. iii. to gr. xx. Some of the special conditions in which this drug has been used are diabetes insipidus, glycosuria, oxaluria, phos-phaturia, nephralgia, neuralgia, hypertrophy of the pros-tate, pertussis, post-partem uterine colic, epistaxis, loco-motor ataxia, epilepsy, hemorrhoids, hemicrania, chorea, morphinism, incontinence of urine in children. It may be given by the mouth, hypodermatically, in a clysma, or in suppository.

2

For its hæmostatic effect it is best given in four-per-cent. solution. For hemorrhoids it may be given with cocaine in suppository to relieve rectal pain. It is best administered in a little water or with a little aromatic syrup. Its incompatibles are said to be many, but sweet spirits of nitre, chloride of iron, and calomel are most to be cautioned against. Fieux states that this drug is eliminated in the milk of nursing women five to eight hours after administration. Antipyrin is a tricky drug, and may be followed by cyanosis, sweating, prostration, and collapse. It is frequently followed by an eruption.

Test (Fieux).—Add 2.5 grams of sodii metaphosphoric acid and 12 drops of sulphuric acid to the suspected fluid (milk, for example), filter, and add a few drops of sodium nitrate to the clear filtrate. If antipyrin is present a clear green color develops.

Antivenin.—A serum perfected by Calmette by injecting cobra venom mixed with solutions of calcium hypochlorite into horses. It is used in doses of 10 to 20 c.c. in severe and dangerous bites of venomous serpents.

Anuria.—Suppression of urine. Seen in uræmia, cholera Asiatica, severe gastro-enteritis of children, and in poisoning by arsenic and oxalic acid.

Anusol (Iodo-resorcin-sulphonate of Bismuth).—Mostly used in the form of suppositories for the treatment of hemorrhoids. Also useful in rectal tenesmus and pruritus vaginæ. Dose is gr. x.

Anytin.—A compound containing carbonic acid, oxygen,

and hydrogen, with 16.5 per cent of sulphur and 4.5 per cent of ammonia.

A brownish-black hygroscopic powder, soluble in water and having the power to make soluble substances which are not so under ordinary conditions, and when used in combination with disinfectants is said to bring out their full antiseptic power.

Aphasia.—Loss of power of speech due to a lesion in the cortex.

Apharsia.—Synonymous with aphasia.

Apical Dulness.—In the differential diagnosis of phthisis, it is to be remembered that Kernig has called attention to the possibility of pronounced dulness at the apex in extremely marasmic patients, at whose autopsies no pathological changes could be found. The dulness of marasmus is always bilateral and accompanied by no other auscultatory signs aside from diminished breath sounds. The dulness may disappear and phthisical symptoms are absent. In some cases the dulness may proceed partly from senile atrophy of the lung.

Apocodeine Hydrochlorate. — An emetic given in croup, pertussis, etc. Dose, gr. ¼ hypodermatically, or ℔ viii. to xx. of a two-per-cent aqueous solution. Also used in chronic bronchitis as an expectorant and sedative in doses of gr. iii. per day.

Apocynum (Canadian Hemp).—This diuretic drug is occasionally of great value in cardiac dropsy, producing diuresis after other drugs have failed to act. It is indicated in cases of uncompensated cardiac lesions, in various valvular diseases, arteriosclerosis, and as an "ultima ratio" in certain forms of threatened pulmonary œdema. The dose of the powder is from gr. x. to xx., that of the fluid extract from ♏ v. to xx.

Apolysin.—A combination of phenetidin and citric acid, analogous to citrophen. It possesses slight antipyretic, analgesic, and diuretic properties, and was at one time used in the febrile diseases of children. It is soluble, acts quicker than phenacetin, and has no bad after-effects. Dose, gr. xxx. to xc. daily. It may also be given in suppository in half-gram dose, every two or three hours.

Apraxia.—A symptom consisting in the loss of the memories of an object, thus leading to the inability to recognize the use, nature, or import of various substances.

Aprosexia.—A term first used by Guye, of Amsterdam, to designate an inability to fix the attention for any length of time. This is frequently seen in conditions of adenoids and in those giving rise to mouth-breathing.

Arbutin.—A glucoside from Arctostaphylos uva ursi. A white powder soluble in water (8 parts) and alcohol. Used in doses of gr. xv., three or four times daily, for catarrh of the genito-urinary tract, especially cystitis due to enlarged prostate.

Argentamin.—This is an amido compound of a silver salt. Has the same properties and use as the nitrate of silver, and is said to be less irritating.

Argonin.—Also known as silver casein. Used in from two- to ten-per-cent solution as a bactericide in gonorrhœa.

At times it dries up the secretion without producing the caustic action of nitrate of silver. It is soluble in albumin in the proportion of 1 to 10 with the aid of gentle heat.

Argyll-Robertson Pupil.—Loss of pupil reflex to light, with preservation of reaction to accommodation. Observed in tabes dorsalis, general paresis, epilepsy (during the attack), and occasionally in diabetes.

Argyria.—A term applied to staining following the continued use of silver preparations. The long-continued administration of the drug internally or its local use may be followed by argyria. Lately a number of instances of conjunctival staining have been reported following the employment of protargol. Argonin, itrol, and actol may give rise to similar conditions. Workers in silver may show an argyrosis of the hands.

Aristol.—A reddish-brown powder composed of iodine (45.8 per cent) combined with thymol, with a pleasant odor, insoluble in water and glycerin. First used in dermatology, in which field it is to-day extensively employed. It has largely replaced iodoform, and in many instances appears quite as efficacious. In indolent ulcers and granulating surfaces it is of great value. It has been used in eczema, lupus, psoriasis, and burns. It may be given in the form of powder, oil, or ointment with lanolin or vaseline in five-per-cent strength; aristol-collodion may also be used.

Arsenauro and Mercauro.—The former is a double bromide of gold and arsenic, the latter a double bromide of gold and arsenic with the addition of mercuric bromide. These are watery solutions which keep well and are readily assimilated. They are said to be excellent alteratives and tonics. Dose, ℔ v. to xv. well diluted after meals.

Asaprol (Calcium-beta-naphthol-sulphonate).—Used in tuberculosis and rheumatism. Analgesic, antipyretic. Dose gr. v. to xv.

Aseptol.—A liquid of syrupy consistence, mixing with water, alcohol, and glycerin. It is obtained by the action of carbolic acid on sulphuric acid. This is used as a dressing for wounds in three- to four-per-cent solution. As a test for albumin in the urine it gives a delicate reaction. Used mostly in surgery.

Aseptolin.—This is a solution containing about three per cent of absolute phenol, and one per cent of pilocarpine-phenyl-hydroxide. The latter is a recent pilocarpine salt. It is given mainly hypodermatically, but may be used per rectum. It is said to enhance the natural antiseptic properties of the blood, and has been used in malaria, influenza, septicæmia. Its main field of application has been in phthisis, in which disease great improvement has been reported. It is not a cure for pulmonary tuberculosis.

Asiatic Pill.—Used quite extensively in the treatment of psoriasis, lichen planus, and other skin diseases. It consists of:

℞ Arsenous acid .... gr. $\frac{1}{8}$.
Powd. black pepper .... gr. i.
Powd. acacia .... gr. $\frac{1}{4}$.
Powd. althæa .... gr. $\frac{3}{10}$.

Or

℞ Acidi arsenosi............................... gr. lxvi.
  Pulv. pip. nigræ........................... ℥ ix.
  Gummi arabici,
  Aquæ. ....āā q.s.
  Div. in pil. No. dccc. S. One to three pills a day after meals.

The various Asiatic pills contain different quantities of arsenic, so that the number to be taken daily will depend upon the amount of the drug present.

Asparagin.—A diuretic, used in cardiac affections, gout, and general dropsy. Dose, gr. i.

Asparolin.—A combination of guaiac, asparagus, parsley, black haw, and henbane, forming a brown liquid. It has been used, in doses of from two to four drachms given in hot water, as an antispasmodic uterine tonic.

Astasia.—A term first used by Charcot to designate motor incoördination in standing, with retention of muscular power and sensation. The patient may be able to move the limbs when in bed, but as soon as he attempts to stand his legs seem to give way under him, or, as Knapp puts it, "bend under him as if made of cotton." This condition in most instances is part of a functional neurosis.

Astraphobia.—Fear of thunder and lightning.

Atheroma.—A term applied to a sebaceous cyst or an encysted tumor, but more frequently to an entirely different condition, viz., a fatty or calcareous degeneration of the arterial walls. The latter is found in lead poisoning, alcoholism, gout, and syphilis; it is part of the senile changes, is hereditary in some families, and occasionally goes with chronic disease of the viscera.

Athetosis.—A condition first described by Hammond and observed principally in the hemiplegia of children. It consists of involuntary and somewhat rhythmical spasmodic movements of the paralyzed parts, especially the fingers and toes. In rare instances the muscles of the face and neck are involved. Motions seem to increase the movements. The term restless ataxia has also been applied to this condition. Strümpell (vol. iii., 1896, p. 551) has an excellent illustration of this affection.

Atmocausia.—Another term for vaporization.

Azoospermia.—The absence of spermatozoids in the seminal fluid.

Azoturia.—An increase of the nitrogenous elements (urea) in the urine.

Baccelli's Sign.—Whispering voice is heard through a serous pleural exudate but not through a purulent one (Guido Baccelli, Rome, 1832).

Bacterial Phosphorescence.—The property possessed by certain microbes of producing light is well known, as in the sea-water organism. Cohn has described a micrococcus phosphorescens, called by Hermes bacterium phosphorescens. The difference in description and name may be explained by an observation of Herman, in which a luminous microbe appeared in cultures at times absolutely spherical (*micrococcus*), while in older cultures it gradually became elongated and rod-shaped (*bacterium*). Cohn's micrococcus inoculated on herring jelly gives a growth which in light is opal color, but in darkness gives out a beautiful, luminous green color.

Bagot's Local Anæsthesia.—This is a combination of cocaine and sparteine, the latter obviating the depressing effect of the former on the heart and at the same time rendering the anæsthesia more lasting. He uses a powder consisting of cocaine hydrochlorate .04, and sparteine sulphate .05; this is dissolved in 1 or 2 c.c. of boiled water. As much as 8 to 10 cgm. of cocaine may be injected without harm.

Bananina.—A fanciful name for a plantain flower, consisting almost entirely of starch. Introduced as a food. It contains phosphates and has a pleasant taste.

Barkoo-Rot.—A skin disease seen in Western Australia. It affects the hands of miners who come in contact with instruments made of some special kind of wood.

Barlow's Disease (Infantile Scurvy).—An affection in which the skeleton is the seat of painful subperiosteal hemorrhages, especially near the epiphyses, associated with profound anæmia, pseudo-paralysis, sponginess or swelling of the gums, and hemorrhages into the skin and mucous membranes. An improper dietary, particularly the prolonged use of proprietary food preparations, has been considered important in the etiology of the disease.

Basham's Mixture (Liquor Ferri et Ammonii Acetatis).—It contains:

℞ Tinct. ferri chloridi.............................. 2
Acid. acet. diluti................................. 3
Liq. ammon. acetatis.... 20
Elixir aromat.,
Syrupi,
Aquæ........................... .. āā q s ad 100
The dose is ℨ ss.-i.

Useful when the combined effect of iron and a diuretic is required. Often successful in erysipelas.

Bassorin Paste.—A quickly drying, unirritating, and readily removable base for the incorporation of remedies used in dermatology, introduced by Lascar and Elliot of New York. It is made from bassorin—a derivative of Bassora tragacanth or other gums—mixed with water, glycerin, and dextrin, in such proportions as to form a jelly resembling vaseline.

Bednar's Aphthæ.—A form of stomatitis seen during the first few weeks of life, and consisting of a bilateral superficial ulceration, of round or oval form and whitish color, which is situated at the posterior alveolar border of the hard palate, at that point where the pterygo-maxillary ligament is put on the stretch during the process of nursing or of opening the mouth.

Benzanilid.—A crystalline white powder used as an antipyretic, especially in infantile practice. Dose for infants, gr. i. to vi. three or four times a day.

Benzoinol.—A petroleum preparation similar to albolene, and used as a vehicle for various medicaments such as menthol, camphor, thymol, eucalyptol, mercury, gaultheria, etc. It is said to consist of albolene with gum benzoin in solution, and is used principally in nose and throat practice.

Benzo-Naphthol.—A white powder, insoluble in water; a powerful diuretic and internal antiseptic, breaking up in the intestine into naphthol and benzoic acid. Dose, gr. v. to x. three times a day.

Benzosol (Guaiacol Benzoate).—The ethereal salt of

guaiacol and benzoic acid. It is a colorless crystalline powder, almost entirely free from odor and taste, insoluble in water, but readily so in alcohol and ether. It is used in phthisis, as a substitute for guaiacol and creosote, as an intestinal antiseptic, and has been recommended in diabetes. The dose is four grains, gradually increased to ten, three times a day, after meals. Piatkowski and J. Blake White have recommended it for diabetes mellitus, in which condition it is said to lower the specific gravity of the urine and control the excretion of sugar.

Beri-Beri.—A disease prevalent in the West Indies, India, Japan, China, Brazil, the Red Sea, and characterized by anæmia, weakness, dyspnœa, œdema, and multiple neuritis. The disease is in all likelihood microbic in origin, although some authors include an excessive use of rice and fish among the etiological factors. The œdema which is often part of the disease has as its favorite sites the sternum and vertebral column.

Bernays' Aseptic Sponge.—So named after Dr. A. C. Bernays of St. Louis. It is composed of properly prepared cotton fibre which is subjected to great pressure. It is presented for use in the shape of compressed circular discs about one-fifteenth of an inch in thickness and of two sizes; a smaller, one and one-fourth inches in diameter, and a larger, one and one-half inches. Their great advantage lies in their great absorptive power; when placed in water they increase in size twelve to fifteen times. This property enables them to exert great pressure when used in small cavities. Dr. W. H. Simpson speaks highly of them as controlling agents in intra-nasal and post-nasal hemorrhage.

Betol.—A white powder, insoluble in water, breaking up in the digestive tract into its constituents naphthol and salicylic acid. Hence it is recommended especially as an intestinal antiseptic. It is also antipyretic in microbic diseases of the intestine, in articular rheumatism, catarrh of the bladder, etc. Dose, gr. iii. to viii. in capsule form. May also be given in suspension.

Caution.—Watch action upon kidney, though this drug is considered less dangerous than the salicylates.

Bier Method in chronic rheumatic arthritides and localized tuberculoses consists in creating a venous stasis by the application of tight bandaging on either side of the articulation or affected region by means of rubber bandages. The skin is to be properly protected and the bandages changed every twelve hours.

Bismal.—Prepared by acting upon the oxide of bismuth with methylenedigallic acid. A powerful astringent used in chronic diarrhœa, especially tuberculous. Dose, gr. ¼ to v.

Bismuth Beta-Naphthol (Orphol).—This is a brown, odorless powder insoluble in water and split up in the intestine into its component parts. It is a powerful and useful intestinal antiseptic in doses of from gr. v. to x. three times a day.

Bismuth Oxyiodopyrogallate.—A fine, yellowish-red, amorphous powder, which is insoluble in water. It is unaffected by the air and light. It is used as an antiseptic for wounds, and is said to be decomposed less readily than other similar preparations.

Bismuth Peptonate.—A brown powder. Dose, ℨi.

Bismuth Subiodide or Oxyiodide.—A brick-red, odorless, tasteless powder, insoluble in water and alcohol. Used with excellent results in the form of powder or ointment as a stimulant to sluggish ulcers.

Black-Water Fever.—A bilious hæmoglobinuric fever seen throughout tropical Africa. It occurs mostly in natives who have had repeated attacks of ordinary fever, or in those who have been in the country for a considerable time; the newcomer is rarely attacked. The clinical features are those of a severe remittent fever, followed by the passage of dark porter-colored urine. The prognosis is bad, though many patients recover.

Blennostasine.—Said to be a bromine derivative of cinchonidine, and consisting of a yellowish solid, soluble in water, insoluble in ether and chloroform. It can be easily crystallized, the crystals being odorless and of a bitter taste. It has been used in catarrh of the upper air passages. Dose, gr. iii. to v. every two hours in capsules or gelatin-coated pills.

Blue Glass.—Blue glass has recently been used in the study of syphilis to distinguish faint eruptions upon the skin. First advocated by Broca. An ordinary spectacle holder may be fitted with glasses of cobalt, similar to those found in ophthalmoscopic outfits. These should be worn close to the eyes, and the examination is best made on a dull day. The method has for its aim the discovery of eruptions before they are revealed to the naked eye, and to discover traces of a preceding eruption and those undeveloped. Syphilides of this order are now called in the Paris hospitals *roséoles à verre bleu.*

Blue Œdema.—Œdema with a cyanotic state seen in certain paralyses accompanied with pain. Probably often a form of hysterical œdema and susceptible of improvement under hypnotic suggestion. It may be associated with hysterical pemphigus (Audry).

Boas' Reagent.—For the determination of the presence of free hydrochloric acid in the gastric juice:

℞ Resorcin resublim................................... 5
   Sacch. alb.. 3
   Spirit. dil...ad 100

Five or six drops of filtered gastric contents are added to the same quantity of this solution and heated until dry in a porcelain dish or spoon over the flame. If free hydrochloric acid is present a rose-red or cinnabar-red coloration will result. This reagent is not used so extensively as that of Günzburg, nor are its results quite so accurate.

Bolognini's Symptom of Measles.—This consists of a slight crepitation which is elicited by an alternating and gradually increasing pressure on the abdominal wall with the three middle fingers of the right and left hand. It is said to give the feeling as if two raw surfaces were rubbing against each other. Koeppen-Norden states that at times this sign is observed within a limited area, and again over the entire abdominal wall. It is said to be an early prodromal symptom of measles, but is by no means pathognomonic. The same sensation is said to occur in cases of intestinal catarrh, and Koeppen-Norden thinks that the source of the Bolognini symptom is to be found in the fluid, frothy, aerated contents of the intestine.

Bordier and Frenkel's Phenomenon.—A rotation of

the ocular globe from above downward during closure of
the eyelids is said by these physicians (*Le Scalpel*, June
19, 1898) to have considerable prognostic value in peripheral facial paralysis. It is observed in grave cases.

Bonain's Local Anæsthesia.—This refers to a local
anæsthesia of the external surface of the tympanic membranes by means of:

    ℞ Phenol,
      Menthol,
      Cocaine hydrochlor...........................āā 1.0
Or
    ℞ Phenol... 2.0
      Menthol.. 0.5
      Cocaine hydrochlor............................. 1.0

Boricin.—A combination of biborate of soda and boric
acid in equal parts. Used as an antiseptic on mucous
membranes, having no caustic or irritative qualities. In
injection or irrigation, ℥ i. to ℥ v. to Oi. water.

Boroglyceride (Glyceritum Boroglycerini).—This contains boric acid and glycerin heated together. In fifty-per-
cent solution it is very efficacious as a depleting agent in
pelvic inflammation.

Botryo-Mycosis.—A condition described by Drs. Pon-
cet and Dor, which consists of an ulcerated fungoid mass
varying in size from a pea to a nut and originating in the
cutis; it resembles granulation tissue, is rich in blood-vessels, and is attached to the underlying tissue by a pedicle,
the entire mass giving the impression of a fungous growth.
In this growth botryomyces are distinguishable, hence the
term botryo-mycosis. This disease is known to occur in

horses after castration. In the four cases observed in man, the finger was affected twice, the thenar eminence and shoulder once.

Bottini's Operation.—The use of galvano-cautery for removing prostatic obstruction by multiple division. First advocated in 1885, revived and improved by Freudenberg. The electrode originally advised had a short beak projecting at an angle from the shaft; after introduction the point is turned downward and withdrawn until it comes in contact with the obstruction; the current is then turned on while a stream of water circulates through the hollow catheter to prevent overheating. Before withdrawal, it is pushed again into the bladder and allowed to cool thoroughly.

Rydygier says that there is danger of death from sepsis, especially when the middle lobe is not enlarged. Frisch has pointed out the necessity of ascertaining which lobe is enlarged before operation; this, however, is not an easy matter. The Freudenberg modification with amperemeter attachment is preferred to the original instrument. Authorities differ as to the advisability of introducing a hard permanent catheter after the operation.

Bouveret Sign in Intestinal Obstruction.—This applies only to the large gut. The cæcum is greatly distended and a large elevation exists in the right iliac fossa.

Bremer's Test for Diabetic Blood.—A greenish-yellow tint developed in the red blood corpuscles after staining with eosin for from six to ten minutes in an oven at 35° C. Distinguished from the ordinary reaction of blood, which gives a brownish color.

Bremer's Test for Diabetic Urine.—Ten cubic centimetres of normal and diabetic urine are placed in two test-tubes, and a small pinch of gentian violet B (Merck) is placed on the surface of each specimen of urine in such a manner as to avoid any of the powder touching the sides of the tubes. In the normal urine the violet floats on the surface, little threads being given off from below which disappear on slight agitation. Diabetic urine is colored within a few seconds from above downward a blue or bluish-violet, and shaking does not produce disappearance of the color. The greater the quantity of sugar, the more rapidly and intensely does the phenomenon appear.

Bromalin (Bromethylformin).—White crystalline laminæ, tasteless and soluble in water. A combination of bromide and intestinal antiseptics, which is said to prevent the occurrence of bromism. The dose is 6 to 12 gm. per day. It is mainly used in the treatment of epilepsy.

Bromidia.—A proprietary preparation which is said to contain in each teaspoonful—

R Chloral,
  Bromide of potash.........................āā gr. xv.
  Extr. cannabis indica,
  Extr. hyoscyamus ....āā gr. ⅛
  Oil of anise, gum arabic, water, and syrup.

It is used rather extensively by the profession, particularly in cases in which it is undesirable to let the patients know what they are getting. Its indications are evident from its composition.

Bromoform.—A colorless, heavy liquid made by the action of bromine and alcohol in the presence of an

3

alkali. The odor and taste are similar to chloroform. This drug has been used as an anæsthetic, but it is unsafe. At present its use is limited to the treatment of whooping-cough, in which disase it sometimes diminishes the number and severity of the paroxysms, but does not shorten the duration of the disease.

The daily dose for an adult is twenty drops; for a child, as many drops as the child is years old three times a day, in solution or emulsion.

Instances of poisoning have lately been reported.

Bromol.—Phenic acid saturated with bromine. Advocated by Rademaker, of Louisville, because of its antiseptic properties (in the treatment of diphtheria), and as a dressing for ulcers. Dose, gr. $\frac{1}{6}$. As an ointment, 3 i. to 3 i.

Bruit du Diable (Venous Hum).—A continuous hum heard in the veins of the neck in cases of anæmia and chlorosis.

Buhl's Disease (Fatty Degeneration of the New-Born). —This disease presents as its most important symptoms a dark coloration of the skin, dark stools, diphtheritic inflammation of the mucous membranes, hæmaturia. A short time after birth symptoms of asphyxia and melæna may appear. The prognosis is bad, death being the usual sequence.

Bulbar Paralysis.—A chronic and slowly progressing paralysis seen in the final stages of chronic muscular atrophy, with atrophy of the nuclei of the twelfth, seventh, ninth, and eleventh cranial nerves. The disease is also known as glosso-labio-laryngeal paralysis; there is disturbance of speech, with tremor and atrophy of the tongue, difficulty in chewing and swallowing; the muscles of the lips and face are weak, rendering whistling and facial expression difficult; there is an increase in the production of saliva, the secretion literally pouring from the corners of the mouth. As a result of paralysis of the ninth and eleventh nerves there is regurgitation of food, loss of pharyngeal and laryngeal reflex, inability to cough. The duration of the disease is from one to ten years, death being due to foreign-body pneumonia, heart failure, or inanition.

Bulimia (Also termed Cynorexia; Heisshunger of the Germans).—An excessive, ravenous, morbid hunger. It may exist as a separate neurosis or as part of an organic affection. Thus it has been observed in Basedow's disease, gastric ulcer with hyperacidity, *tænia*, diarrhœa, menorrhagia, hysteria, *neurasthenia*, hypochondriasis, psychoses, *pregnancy*, tuberculosis, *diabetes*.

This condition is also occasionally observed in gastric carcinoma, dilatation of the stomach, cerebral syphilis.

Buphthalmia.—Also known as kerato-globus, and signifying distention and protrusion of the cornea.

Burma Head.—A well-known malady of Rangoon and other regions in the Burmese territory. In the acute stage there are loss of memory, signs of idiocy, homicidal mania, together with an inability to walk. The etiology is said to be at times a too free use of spirits.

Butter-Cyst.—The name applied to cystic tumors (of the breast) having semi-solid contents of yellowish-brown color and of buttery consistence. The contents may harden on exposure to the air. Probably often a galactocele whose watery constituents have become absorbed.

Cachexia Strumipriva.—A condition resembling myxœdema and following the total extirpation of the thyroid gland.

Cacodylic Acid.—Recommended by Jockleim as a succedaneum for arsenical preparations, being rich in arsenious acid (fifty-four per cent, Danlos). Occurs as rhombic prisms, soluble in water.

Recommended for psoriasis:

℞ Acid. cacodylic.................................... 5
   Spir. sacchari,
   Syr. aurant. cort...............................āā 40
   Ol. menth. pip................................... q.s.
   Aquœ dest.. 120

Cadmium Salicylate.—Is said to be more energetic as an antiseptic than other cadmium salts. Used in purulent ophthalmia, conjunctivitis, gonorrhœa, etc.

Caisson Disease.—An illness suffered by workers in compressed air, as in tunnel boring, mining, and bridge building. A paralysis succeeds to painful sensations in the limbs, especially about the knees.

Camphoid.—A substitute for collodion, consisting of—

℞ Pyroxylin.. 1
   Camphor,
   Dilute alcohol...................................āā 20

Camphoric Acid.—Derived from camphor by oxidation with concentrated nitric acid. Colorless crystals, practically insoluble in water, but soluble in ether, hot water, and alcohol.

Used principally in doses of gr. xv. for the night sweats of phthisis; in from one-half to three-per-cent solutions used as a spray and gargle.

Camphor-Phenol.—Camphor dissolved in three parts of ninety-five-per-cent carbolic acid. Used externally in a fifty-per-cent oily solution. It is antiseptic, anæsthetic, antipruritic. Internally carminative. Dose, gtt. v. to x. in capsule.

Camphor-Salol.—Salol three parts, camphor two parts. Used as an external antiseptic.

Camphor-Thymol.—A fusion of equal parts of camphor and thymol, giving an oily and insoluble substance; it is non-irritating but less active than mentho-phenol.

Cancroin.—A solution of neurin in water with carbolic and citric acids. Used by Adamkiewicz as a prophylactic against carcinoma.

Cannabinol.—A preparation of cannabis indica. A hypnotic which is said to be more reliable than cannabis indica itself. It is rather a sleep producer than a sleep forcer.

Captol.—This new antiseborrheal agent and hair cosmetic is a condensation product of tannin and chloral, and forms a dark-brown hygroscopic powder, which is hardly soluble in cold water, but soluble in alcohol and warm water. A one- to two-per-cent alcoholic solution, twice daily, was found by P. J. Eichhoff to be of excellent service in seborrhœa capillitii and the falling out of the hair dependent upon it. No injurious effects were ever noticed.

In eight to fourteen days the hair ceased to fall out. As a prophylactic a captol solution has also been recommended.

Now on the market only as a proprietary remedy.

Carbazotic Acid.—Synonymous with picric acid.

Carbonic Snow. See Crymotherapy.

Cardioscope.—An instrument devised by Czermak to facilitate the investigation of heart movement in animals, the contractions being reflected by means of mirrors.

Cardol (Tri-bromo-salol).—Said to have narcotic and hæmostatic action, with some hypnotic powers.

To produce sleep, ℈ ss., subsequently reduced to gr. xv.

Carnogen.—See Red Bone Marrow.

Caroid.—Said by Chittenden to be a true soluble digestive ferment of vegetable origin. It is derived from the Carica papaya, and consists of a pale yellow, dry, non-hygroscopic powder. It is said to convert all forms of albumin into peptone, whether the medium be alkaline, acid, or neutral. It has been used in diminished secretion of gastric juice, atonic dyspepsia, achlorhydria, carcinoma, gastralgia, intestinal fermentation, and infantile indigestion.

Dose, gr. i. to iii., administered in powder or tablet form as such, or with charcoal, bicarbonate of sodium, boric acid, etc.

Cassareep, or Cassaripe.—From the juice of a South American plant, cassava, recently introduced by Dr. H. B. Chandler, of Boston, and advocated also by Dr. S. D. Risley (*Philadelphia Medical Journal*, October 29, 1898) in corneal ulcers and other infectious diseases of the eye. Used in ointment up to ten-per-cent strength.

Castoria.—This is said to have the following composition:

℞ Senna	℈ iv.
Manna	℥ i.
Rochelle salts	℥ i.
Fennel, bruised	℈ iss.
Boiling water	℥ viij.
Sugar	℥ viij.
Oil of wintergreen	q. s.

Pour the water on the ingredients. Cover and macerate until cool; strain and add the sugar, dissolve by agitation, and add oil of wintergreen to flavor.

Cearin.—Carnauba wax and ceresin one part, liquid paraffin four parts. An excipient for chemical salts which decompose in the presence of lard. Used in making ointments of iodide of potassium, acetate of lead, etc.

Celloidin.—A very concentrated collodion used as a protective dressing in wounds, etc.

Cellular Therapy.—The definition given by Aulde is: "The method in therapeutics of exhibiting properly selected medicaments with a view to restoration of cell-function. It aims to supply scientifically those remedies that experience has shown to possess special curative properties in the restoration of disordered functions."

Cerebrin.—A preparation from the gray matter of the brain (sheep, calves) made with equal parts of glycerin and one half-per-cent carbolic-acid solution. Has been used in chorea, locomotor ataxia, etc. Dose, gtt. v. to xx.

This term is also applied to an elixir containing analgesin, cocaine, caffeine, and ether. It is said to be an active anti-neuralgic. Dose, two to four teaspoonfuls.

Cerebro-Spinal Sclerosis (Multiple Sclerosis).—A cerebro-spinal affection of which the classical symptoms are scanning speech, nystagmus, intention tremor, spastic gait, and exaggerated patellar reflex.

Ceyssatite.—An absorbent powder, first brought to the notice of the profession by Veyrières in May, 1898. It is a fossil earth from the village of Ceyssat, and is composed almost entirely of pure silica. According to Darier, its composition is:

Silica	64.570
Water	7.130
Chalk	2.850
Iron	5.650
Magnesia	6.480
Organic matters	7.330

It is principally used as an absorbent dusting-powder in intertrigo, eczema, and hyperidrosis. It may be applied in ointment form.

Charcot's Joints.—An affection of the joints observed in locomotor ataxia and characterized by a rapid, painless swelling and enlargement, with subsequent extreme mobility and grating. The effusion into the joint is excessive, and the knees are most frequently involved.

Chautard's Test.—This is a test for acetone in the urine. Make a solution of fuchsin (1:2,000) and decolorize with sulphuric acid; pour half an ounce of urine into a test-tube and add a few drops of the fuchsin solution; in the presence of acetone the urine assumes a violet or purple color, the depth being proportionate to the quantity of acetone.

Cheiragra.—In contradistinction to podagra this term is applied to the involvement of the joints of the hand in acute gout.

Cheiromegalia.—A localized hypertrophy of the hands coming on in syringo-myelia.

Chenopodium Ambrosioides (American Wormseed or Jerusalem Bark).—This drug belongs to the class of anthelmintics. The dose of the oil is ℥ii. to viii. for a child two years of age.

Cheyne-Stokes Respiration.—A form of dyspnœa with gradual and rhythmical increase of respirations, followed by a temporary arrest and again by an increase.

Observed most frequently in uræmia, and occasionally in cardiac and cerebral diseases. It is as a rule of evil significance, though cases of recovery have been reported. In rare instances it is seen in normal persons during sleep. —JOHN CHEYNE, London (1777–1836).

Chinaphtol (Beta-naphtol and Monosulphate of Euquinine).—A yellowish, bitter, insoluble powder. It has been used in syphilis, dysentery, intestinal tuberculosis, and especially in acute articular rheumatism. Dose, gr. viii. in capsule four or five times a day.

Chinosol (Quinosol).—A succedaneum of corrosive sublimate and carbolic acid, said to be free from toxic effect and irritation. An antiseptic and deodorant proprietary preparation, used in the form of powder or solution (1 : 500) for ulcers, vaginal douches, disinfection of excreta.

Chloralamide (Formidate of Chloral, introduced by

Von Mering).--It consists of clear, white, brilliant crystals of a bitter, unpleasant, pungent taste, partly soluble in water, but more so in alcohol. It is an hypnotic, its action depending upon the slow elimination of chloral by the alkaline blood. It has 'proven of value in insomnia, and Peabody has recommended it for phthisis, rheumatism, alcoholism, and delirium tremens. The dose is gr. xxx. to xlv., in powder form or dissolved in spirits or wine. It is a soporific pure and simple, and is practically useless in the insomnias due to pain. The effects begin in half an hour and last up to eight or nine hours.

Chloralimide.—Colorless crystals, insoluble in water. Action and dose those of chloral.

Chloralose.—A hypnotic, giving refreshing sleep without bad effects. Dose, gr. iv. to xii. at bedtime.

Chlorhydria.—An excessive amount of hydrochloric acid in the stomach.

Test.—Let the patient swallow a small quantity of saturated solution of sodium bicarbonate while auscultation over the stomach is carried out. If normal hydrochloric acidity is present, a fine crepitation is heard. In superacidity, the sound is louder and occurs sooner. In anacidity, crepitation is absent.—BENEDICT.

Chloride of Tin.—A saturated solution of stannous chloride in distilled water is said to clean rusty instruments.

Chlorobrom.—A combination containing chloralamide and potassium bromide, gr. xxx. each to the ounce, flavored with licorice. Recommended by Hutcheson for sea-sickness in dose of ℨ iii.

Chlorophenol.—An antiseptic especially recommended in tuberculosis of the larynx and tongue, in five- to twenty-per-cent solution. Also recommended for lupus vulgaris (Barbe).

Chromidrosis.—Colored sweat.

Chvostek Phenomenon of Tetany.—Striking the cheek with the fingers or a percussion hammer gives rise to marked contraction of the facial muscles on account of the peculiar nerve irritability which exists in this condition.

Cimolite (A perfumed Kaolin).—Dusting-powder.

Cinnamic Acid.—Recently recommended as an antituberculous agent in medicine, surgery, and dermatology.

Citrurea.—Citric acid, urea, and bromide of lithia. Indications the same as for urea and lithium.

Claustrophobia.—A fear and dread of closed spaces. The person cannot attend the theatre, church, or assembly without a sudden distressing sensation of anxiety and apprehension, for which he can give absolutely no reason and which causes him to seek the open air, when he at once feels completely relieved.

Cling Symptom (Klebe Symptom).—In fæcal accumulation, when decided finger pressure is made upon the tumor-like mass, the intestinal mucous membrane clings for a time to the depression made on its surface and may be subsequently felt to release itself.

Cobalt Nitrate.—This has been said to be a successful

antidote (*Medical Standard*, November, 1898) in poisoning by hydrocyanic acid and potassium cyanide.

Cocaine Cantharidate.—A mixture of cantharidate of soda, with one per cent of hydrochlorate of cocaine. Used hypodermatically in laryngeal tuberculosis, etc.

Cocaine Phenate.—Pure cocaine dissolved in alcohol, to which an alcoholic solution of phenic acid is added up to saturation, and evaporation of the alcohol.

Coin Sign (Fr., Signe du Sou; Ger., Stäbchenplessimeter percussion).—In this form of auscultory percussion a coin or piece of metal placed against the chest opposite to the examiner's ear is struck with another piece of metal. A dull wooden sound is transmitted by healthy lung tissue. Dulness is also heard through tuberculous and pneumonic foci unless infiltration is very excessive. In pneumothorax there is a brazen resonance. Pleurisy with effusion gives a clear sound, and the level of the fluid may be determined.

Colchicine.—The alkaloid and active principle of colchicum autumnale. Used latterly instead of the various preparations from the seed and root. It is a yellowish powder, with a bitter, lasting taste, soluble in water and alcohol. It acts as a diuretic in gout, rheumatism, and Bright's disease. Dose is gr. $\frac{1}{80}$ to $\frac{1}{60}$ three times a day. In doses of gr. $\frac{1}{125}$ hypodermatically it is said to act as a specific in gout.

Coley's Mixture.—This consists of the toxins of the streptococcus erysipelatis and the bacillus prodigiosus, and has been used as a remedy for cancer, especially in the early stage. It has also been recommended to alleviate the symptoms in those cases which are too far advanced for surgical interference. It is especially advocated in inoperable sarcomata. Of eighty-four round-celled sarcomata treated by Coley, thirty-five were more or less improved, one remained well upward of three years, one for one and one-half years, and one for one year. Of all cases eight remained well over three years.

Collesin.—A cutaneous dressing introduced by Schiff as a skin varnish.

Colloidal Mercury.—See Hydrargyrum Colloidale.

Corrigan's Pulse.—The water hammer, jerking, or collapsing pulse of aortic regurgitation. Also known as pulsus celer et altus.—DOMINICK JOHN CORRIGAN, Dublin (1802–1880.)

Cotarnine Hydrochlorate.—See Stypticin.

Coxa Vara.—Deflection or bending of the neck of the femur.

Cracked-Pot Sound.—A recently advanced diagnostic sign in cerebellar tumor. Due to a separation of the sutures and hence found only in those cases in which the sutures are widely separated. It is also present in extensive linear fractures of the skull. Best brought out by digital percussion without the pleximeter.

Craniometer.—An instrument for measuring the skull has been introduced by Krönlein. It is a simple contrivance, consisting of six thin and narrow strips of flexible metal, each adjustable on a slit screw. It is illustrated and described in the *Centralblatt für Chirurgie*, January 7, 1899.

Credé's Method.—A prophylactic measure against ophthalmia neonatorum, which consists of the instillation into the eyes of new-born children of a few drops of a one- to two-per-cent solution of nitrate of silver. This method has fairly revolutionized the subject of purulent ophthalmia of the new-born, and has reduced the mortality to a very great extent.

Credé's Method.—Manual expression of the placenta.

Credé's Ointment.—A soluble silver ointment made from argentum colloidale, suggested by Credé as a cure for septicæmia and pyæmia. It is applied by inunctions, the medicament reaching the infected blood by percutaneous absorption. It is supposed by the formation of powerful bactericidal silver salts to insure universal antisepsis or disinfection of the entire organism.

Dose.— 3 ss. to 3 i. repeated every twelve hours till symptoms abate.

Creolin.—A dark-brown fluid with a tar odor, procured from coal-tar creosote. It is an antiseptic and disinfectant, and said to be less poisonous and more active than carbolic acid. It emulsifies with water and is used in one-half to one-per-cent solution as a douche. It has been used in cystitis, foul-smelling ulcers, bone necroses, and pityriasis versicolor.

Creosoform.—A greenish powder formed by combining formic aldehyde with creosote.

Creosol (Homo-pyro-catechin-mono-methyl-ether). — An antiseptic.

Creosotal (Creosoti Carbonas).—A clear, light-brown, viscous liquid, containing ninety per cent of pure creosote, with a mild flavor and no caustic properties, insoluble in water, but soluble in oil. It may be given in daily doses ranging from five drops to three teaspoonfuls, since it is said not to derange digestion, the creosote being disguised in the neutral combination, thus permitting larger dosage than is possible with creosote alone. Especially useful in tuberculosis. In infants, ℞ iv. to viii. Jacob and others report good results from smaller doses (gtt. v. three times daily, increasing by three drops each day up to twenty-five-drop doses; keeping at this dose for several weeks, and then decreasing gradually). It may be given pure in capsules, in emulsion, milk, coffee, wine, or with cod-liver oil.

Creosote Phosphate.—A colorless liquid obtained by Dr. Brissonnet (*Journal des Sciences médicales de Lille,* August 20, 1898) and used in tuberculosis. It is said to increase the amount of urea and the urinary acidity.

Creosote Tannophosphate.—An amber-colored fluid obtained by Dr. Brissonnet and used in phthisis. Action same as phosphate.

Creosote Valerianate. — Recommended by Grawitz for pulmonary tuberculosis. Zinn (*Therap. Monatshefte,* March, 1898) reports eighty cases treated at Gerhardt's clinic. It was used in all forms of tuberculosis without any gastro-intestinal disturbance. Dose, ℔ iii. to iv. in capsules three or four times a day.

Crymotherapy.—A term used by M. Ribard (*Gazette hebdomadaire de Médecine et de Chirurgie*, September 1, 1898) to mean the therapeutical use of great cold, locally applied. He uses a bag filled with "carbonic snow" at a temperature of 176° F. below zero, and applies it to the stomach for half an hour daily. This is surrounded with cotton to prevent injury to the skin. These applications are said greatly to increase the appetite in consumptive patients. The price of carbonic snow makes the suggestion scarcely practical.

Crystallose.—A substance said to be five hundred times sweeter than sugar and free from taste or odor other than that of the pure granulated article.

Curschmann's Spirals.—Found in all true cases of bronchial asthma. These may be seen with the eye, magnifying glass, or, better, with the microscope. They consist of a twisted, spirally arranged mass of mucin, in the centre of which is seen a central thread of much lighter color.—HEINRICH CURSCHMANN, Hamburg, 1846.

Cutol.—An astringent, containing 76 per cent tannin; alumina, 13.23 per cent; boric acid, 10.77 per cent. Soluble cutol is obtained by combining with tartaric acid. A light brown powder, insoluble in water.

Darwinian Tubercle.—One of the supposed stigmata of degeneration, consisting of a small nodule on the margin of the helix, at the apex of the fawn-like ear.

Denisensko's Method in cancer: applications of extract of chelidonum majus.

Dermatol.—This is the subgallate of bismuth. A sulphur-yellow, odorless, tasteless powder, insoluble in water. Used in surgery, dermatology, and gynæcology for burns, ulcers, herpes, hyperidrosis, leucorrhœa, and the like. Internally it is used in gastro-enteritis, dysentery, and diarrhœa. Dose is gr. v. to x. three times a day.

Dermatothlasiomania.—A mania for injuring the skin, epilating hairs, continually expressing comedones, or pinching the face.

Dermography or Dermatography.—A vascular neurosis of the skin which permits of tracing upon the latter with the finger-nails, pencils, or other rather blunt objects; the lines remaining as urticarial wheals. Seen in hysteria, neurasthenia, urticaria, etc.

Diabetin.—A white, dry powder of a sweetish taste used in diabetes. It acts favorably upon the nutrition of diabetic patients without increasing the amount of sugar in the urine or blood. It has been recommended to replace cane sugar, its advantage being that it aids in the formation of tissue and acts as a sweetening agent at the same time. Its great disadvantage is its expense.

Diaphthol.—The same as quinaseptol.

Diazo Reaction (Ehrlich).—A reaction of the urine first introduced by Ehrlich and considered by him diagnostic of typhoid. The test consists of two solutions:

4

I.

Acidi sulphanilic	5.0
Acidi hydrochlor. pur	50.0
Aquæ destillat	100.0

II.

Sodii nitros	0.5
Aquæ destillat	100.0

Take 50 c.c. of **I.** and 1 c.c. of **II.** and add an equal quantity of urine. Now add one-eighth the volume of ammonia and shake thoroughly. If the reaction be positive a rose-red froth will appear; if negative a brownish-yellow froth (Klemperer). This test has failed to sustain all that Ehrlich claimed for it. While corroborative for the diagnosis of typhoid it is by no means pathognomonic. The diazo reaction has been observed in typhoid, pneumonia, measles, miliary tuberculosis, sepsis, and severe cases of phthisis.

Digitoxin.—The glucoside from Digitalis purpurea. White odorless crystals insoluble in water, soluble in chloroform and alcohol. It is a cardiac tonic and diuretic in doses of gr. $\frac{1}{250}$ to gr. $\frac{1}{120}$ two or three times a day.

Dionine.—A new derivative of morphine, to replace the latter or codeine when intolerance for them is found. It is said never to produce headache, vomiting, or constipation. An odorless, crystalline, white powder, soluble in water. It is used in bronchitis, dyspnœa, asthma, etc. Dose, gr. $\frac{1}{3}$ to i. daily.

Diuretin (Theobromine Sodium Salicylate).—Thiss is a white powder, with a sweetish alkaline taste, soluble in half its weight of water. It acts directly upon the kidney epithelium, producing a diuretic action in doses of gr. xv. four or five times daily. The maximum daily dose is ℥ ii. It is indicated in all cases of dropsy due to cardiac or renal affections, and often produces good results when digitalis, caffeine, and strophanthus have failed. It may be administered alone or in combination with heart tonics. It is best administered in capsules or solution.

Dobell's Solution.—

```
℞ Acidi carbolici.............................. gr. x.-Ꝺi.
  Sodii boratis,
  Sodii bicarbonatis.......................āā ℥ i.
  Glycerini ................................. ℥ i.
  Aquæ............... .................... Oi.
```

An alkaline disinfectant solution used extensively in nose and throat practice. The quantity of carbolic acid and alkalies may be varied to suit the exigencies of the case. Listerine may be substituted for glycerin, or rose-water may be used instead of the ordinary water.

Douglass Antiseptic Nasal Tablet.—

```
℞ Trit. sod. chloride..................... (1-1,000) gr. v.
  Trit. sod. bicarb...................... (1-1,000) gr. v.
  Trit. zinc chloride...................(1-10,000) gr. $\frac{5}{10}$
  Trit. mercury bichlor................ (1-100,000) gr. $\frac{5}{100}$
  Oil of wintergreen .......................... ♍ $\frac{1}{20}$
  Tinct. saffron................................ q.s.
     M. ft. tab. No. i.   S. Two tablets in eight ounces of water.
Make a non-irritating, antiseptic, and slightly astringent solution.
```

Duboisine. — Alkaloid from Duboisia myoporoides. Colorless crystals, partly soluble in water, easily so in alcohol, ether, and chloroform. Used as a mydriatic in same strength as atropine. As a sedative and hypnotic in mental diseases and for paralysis agitans, gr. $\frac{1}{200}$ hypodermatically.

Duhring's Disease.—Dermatitis herpetiformis. According to Dr. Duhring, this is an inflammatory, super-ficially seated, multiform, herpetiform eruption, character-ized mainly by erythematous, vesicular, pustular, and bullous lesions, occurring generally in varied combinations, accompanied by burning and itching, pursuing usually a chronic course with a tendency to relapse and to recur.

Duotal.—See Guaiacol Carbonate.

Dural Infusion.—The utilization of the lumbar punc-ture for direct therapeutic applications in cerebrospinal diseases. Antitetanus serum has been thus injected. (JABOULAY.)

Duroziez's Double Murmur.—A double murmur audi-ble on pressure with the stethoscope over the femoral artery in cases of aortic insufficiency. The first murmur is caused by the diastolic blood stream, the second by the returning stream produced by the regurgitation.

Dysæsthesia.—The state in which a sensation is per-ceived a number of seconds after the excitation has been applied.

Dyschromatopsia.—Loss of vision for various colors. Observed in optic atrophy (tabes dorsalis).

Dyscrasia.—A morbid condition of the system.

Dyspareunia.—Excessive pain during coitus.

Eau d'Alibour.—

℞ Water	200 parts.	
Saturated camphor water....................	q.s.	
Copper sulphate............................	2	"
Zinc sulphate..............................	7	"
Saffron....................................	0.4	"

Recommended by Sabouraud for impetigo.

Eberth Bacillus.—The bacillus of typhoid fever. To distinguish it from the coli bacillus, it has been found that the latter will grow in bouillon containing arsenious acid, while the former will not, if even a small percentage is present. It is said that the coli bacillus may be made to grow in media containing as much as three per cent of the acid. This is thought to be a valuable biological reaction.

Echophony.—A form of echo, heard on auscultation, immediately following vocal sounds. A symptom described by Serrand in Woillez's disease.

Ecthol.—Anti-purulent and anti-morbific. Said by the American manufacturers to contain the active principles of echinacia and thuja. Dose, teaspoonful four times a day.

Ectocardia.—A condition in which the heart is displaced. Holt (*Medical News*, December 11, 1897) has reported an exceedingly interesting case in which the organ was covered only by integument and subcutaneous tissue.

Edlefsen's Treatment for Chronic Eczema.—

℞ Pure iodine		0.1
Iodide of potassium		0.25
Glycerin		12.0

M. S. Apply every evening.

Electric-Light Blindness.—A condition somewhat similar to that of snow blindness and the smoke and heat blindness of firemen after a conflagration. It results from exposing the eyes unprotected by glasses to intense and prolonged electric illumination. A wash containing camphor and borax is said to be very efficacious in mild attacks.

Electric-Light Treatment.—Ewald having noted a diminution in the number of instances of rheumatism, neuralgia, migraine, etc., in workmen subjected to the action of electric light, Kozlovski and others have applied this agent therapeutically by means of electric-lighted cabinets or apartments. One method is to place the patient one and one-half metres from the light; he is protected by blue spectacles and by a screen having an aperture through which the light falls upon the particular region of the body to be acted upon. Three or four sittings at intervals of three or four days are said to produce an amelioration, while ten or twelve may be necessary for a cure. A closed cabinet may be used in which the patient lies as in a bath. Considerable cutaneous irritation may sometimes be produced. The term radiant-heat bath is preferred to electric-light bath by Kellogg.

Elixir Chloralamide.—An efficient and relatively safe hypnotic, each tablespoonful containing gr. xv. of chloralamide. It may be given as such or in combination with codeine, morphine, heroin, bromides, etc. It has been used by Professors Dana, Smith, Wilcox, H. C. Wood, etc., with good results.

Ellis' Line.—A parabolic or S-shaped percussion line which is observed in pleurisy with effusion. When the patient is in the erect position the upper border of the dulness is higher behind than in front.

Emol.—Silicious earth containing alumina, lime, and steatite. Ingredient of Emol Keleet, a soothing dusting powder. Emol is said to act as a natural soap.

Emplastrum Salicylatis Saponatum.—This consists of:

℞ Acidi salicylici 1.25
 Emplastr. diachyli,
 Emplastr. saponis .... āā 10.0
 Vaselini 4.0
M.

Used as an external application in boils, carbuncles, abscesses, chronic eczemas, etc. It forms a rather hard ointment which must be warmed, then spread on linen and applied.

Empyema Necessitatis.—An external pulsating empyema.

Enteroclysis.—Washing out the intestines by rectal irrigation by means of a long rubber tube. Indicated in uræmia, shock, etc.

Enteroptosis.—A general term which is applied to prolapse of the abdominal organs—stomach, colon, kidney, spleen, liver, etc.

Enuresis.—Involuntary evacuation of urine, either during the day or night. It may occur as a part of a neurosis, or in conditions of cystitis, stones in the bladder, vulvo-vaginitis, oxyuris vermicularis, ascarides, phimosis, adenoids, masturbation, and epilepsy.

Epicondylalgia.—An occupation neuralgia occurring in the muscular mass about the elbow-joint, following fatiguing work. Treatment by massage is efficacious.

Epidermin.—An artificial covering for denuded skin, supposed to be a combination of white wax and powdered acacia in equal parts thoroughly triturated, to which are added an equal amount of water and glycerin which have first been mixed and brought to the boiling-point.

Epityphlitis.—A term proposed by Küster as a substitute for appendicitis.

Epstein's Pearls.—Small, irregular, slightly elevated, yellowish-white masses, seen on either side of the median line of the hard palate at birth. They have no particular significance and occasionally ulcerate.

Erb's Reaction.—See page 116.

Erben's Pulse Phenomenon.—In neurasthenia, if the patient bends forward or makes an attempt to sit down, a distinct and appreciable bradycardia is observed. The pulse returns to the normal in a very short time whether the patient assumes the erect position or not. This occurs even in the presence of tachycardia. Erben (*Wiener klinische Wochenschrift*, 1898, No. 24) states that in healthy individuals and in tachycardia dependent on conditions other than neurasthenia this phenomenon does not occur.

Eriodictyon.—See Yerba Santa.

Erysipeloid.—An affection of the hands in fishermen and fish-cleaners especially, resembling erysipelas of limited extent, starting often in one finger and invading the others in succession, usually following some slight local injury. Treatment by ichthyol is usually successful.

Erythema Enematogenes.—This term is applied to an eruption (erythematous, scarlatiniform, morbilliform) which sometimes occurs in children after the administration of enemata. This eruption appears twelve to twenty-four hours after the enema and lasts from twenty-four to forty-eight hours. It is usually located on the anterior surfaces of the knees, backs of the elbows, buttocks, and face, is rarely followed by desquamation, and does not give rise to any constitutional disturbance. This eruption is probably due to a vaso-motor disturbance which permits the absorption of the intestinal contents or of the substances contained in the enema.

Erythroblast.—A nucleated red blood cell, whether it be a microblast, normoblast, mesoblast, megaloblast, or gigantoblast. The presence of nucleated red blood cells in the blood is a *signum malum* and is met with in severe cases of chlorosis, in pernicious anæmia, leukæmia, and carcinoma. They are normal constituents of the blood of the new-born.

Erythrol.—The double iodide of bismuth and cinchonine. Used by Robin in butyric fermentative dyspepsia in doses of gr. ¼ after meals. It is given in powder form, generally with magnesia.

Erythromelalgia (Weir Mitchell's Disease).—A condition associated with severe pain in the heel and sole of the foot, with redness, swelling, and pulsation of the vessels. In exceptional cases the affected parts assume a dark red or bluish-red appearance, hence the term " red neuralgia." The condition is occasionally associated with hyperidrosis. Drs. Mitchell and Spiller (*American Journal of the Medical Sciences*, Jan., 1899) have published a case in which peripheral disease of nerves and arteries was demonstrated.

Esbach (Albuminometer and Solution).—This name is associated with the quantitative test for albumin in the urine. The albuminometer resembles a test-tube, but is marked with a series of fine graduations from 1 to 12 and with the letters U and R. The solution used for testing, known as Esbach's reagent, consists of 10 gm. of picric acid,

and 20 gm. of citric acid to the quart of water. Add the
urine to U, the reagent to R, stop with rubber cork, and
turn upside down two or three times. Place the tube in the
erect position for twenty-four hours, and at the end of that
time notice where yellowish-white precipitate stands. If at
1, there is one-tenth of one per cent of albumin; if at 5, one-
half of one per cent, etc. While the results with this apparatus are not accurate, they are approximate enough for
practical purposes.

Etherion.—The name given to a new element which Mr.
Charles F. Brush claims to have discovered in the air. It
is said to have great heat conductivity at low pressures.
Sir William Crookes considers it water vapor.

Ether Pneumonia.—Said by Whitney to be a microbic
disease preventable by thoroughly cleansing the mouth,
nose, and palate of the patient and the mouth-piece employed by the anæsthetist. It is thought by Drummond,
however, that ether is the most important cause. Most
cases follow abdominal operations. It occurs once in
about three hundred cases (ANDERS).

Ethyl Chloride (Kelene).—A very valuable local anæsthetic, especially in non-inflammatory conditions, such as
amputations of fingers and toes, ingrowing toenails, circum-
cision, removal of foreign bodies, glands, and small tumors.
Also useful for the incision of abscesses and felons and in
neuralgia. It is sold in the form of sealed glass tubes holding about 10 grams. By holding the tube at a distance
of four to six inches from the part, the liquid is ejected in
the form of a very fine spray against the part of the skin to
be anæsthetized. Anæsthesia is produced by freezing, in

from ten to fifteen seconds. The ethyl chloride Bengué is the most reliable.

Eucain.—This is a local anæsthetic, first discovered by Dr. Kiesel, of Berlin. Its uses and method of employment are similar to those of cocaine. Its advantages over cocaine are said to be, that it does not cause mydriasis or disturbance of accommodation, the solutions keep better and are not altered by sterilization, there is less danger of poisoning, and its after-effects subside more quickly.

The solutions used vary from two to ten per cent. One cubic centimetre of a two-per-cent solution may be used hypodermically. A mixture of cocaine and eucain would seem to possess an advantage over either substance alone.

Alpha-Eucain is used in surgery of the nose, etc.; beta-eucain also in dentistry, ophthalmology, etc. The latter is said to be nearly four times less toxic than cocaine.

Eucasin.—Casein treated with ammonia so as to render it readily soluble. It is a food and is readily taken in the form of cakes, thick soup, and chocolate. It is said to be especially valuable in chronic phthisis (Hugo Weiss), gastric ulcer, atony of the stomach, chronic intestinal catarrh, anæmia, and chlorosis.

Euchlorine.—

℞ Potassium chlorate	gr. xviij.-xxx.
Hydrochloric acid	gtt. x.
Water	℥ viij.

Said to have been used with benefit in diphtheria, as gargle, spray, etc.

Eudermol.—This is nicotine salicylate containing fifty-four per cent of nicotine.   Used in ointment form for ring-worm.

Eudoxin.—The bismuth salt of nosophen. Has the same properties as dermatol, aristol, airol, etc. Dose, gr. v. internally three times a day as an astringent and anti-septic.

Eugallol (Pyrogallol Monoacetate).—This is a syrupy, transparent, dark yellow mass easily soluble in water. It is very useful in chronic and stubborn cases of chronic eczema and psoriasis.   It is best used in the form of a paint smeared over the patch.   This drug is very rapid and vig-orous in action, and must be carefully watched.

Eulexin (Aromatic).—A proprietary preparation used for diabetes mellitus, and said to consist of jambul, Para-guay tea, cascara sagrada, aromatics, and glycerin.

Eunatrol.—Oleate of soda, a white powder, cholagogue. Dose 3 ss. to 3 i. per diem in capsules.

Euphorin (Carbamate of Ethyl and Phenyl).—A white crystalline powder slightly soluble in water.   Antipyretic, antiseptic, antirheumatic.   Initial dose, gr. iss.; maxi-mum, gr. xv.

Euphthalmine Hydrochloride.—A powerful mydriatic in five- to ten-per-cent solution. Said to be superior to homatropine.

Euquinine.—An ethyl carbonic-acid ester of quinine. It occurs in white crystals, readily soluble in alcohol, ether, and chloroform, but with difficulty in water.   The alkaloid itself is tasteless, with a bitter after-taste.   It is best given

in sherry, milk, soup, or cocoa. The dose is gr. xv. to
xxx. daily. It is especially useful in infantile practice.
This drug has the same remedial virtues as quinine without
its unpleasant properties, and may be used in all cases where
quinine is indicated. There are a chloride, sulphate, and
tannate of the drug, their degrees of solubility being given
in the order of their arrangement. The hydrochlorate is
bitter and possesses no advantages over quinine.

Eurobin.—A compound of chrysarobin and acetic acid.

Europhen.—Produced by the action of iodine on iso-
butylorthocresol, and consisting of an antiseptic, amorphous, yellowish powder containing about twenty-eight
per cent of iodine. Its properties are those of iodoform,
but it is less poisonous and has a less disagreeable odor.
Used for wounds, burns, and ulcers in the form of powder,
or with olive oil, lanolin, collodion, or traumaticin. Very
serviceable in ulcus molle.

Do not mix this powder with starch, zinc oxide, mercuric
oxide, or the salts of mercury, because decomposition will
occur.

Used externally in ointment form in strengths of
from two to ten per cent; subcutaneously, gr. i. to ii.

Exalgin (Methylacetanilid).—A nerve sedative in daily
doses of gr. iv. to v. Given in capsule or alcoholic mixtures. Also analgesic and antithermic.

Exophthalmic Goitre (Also known as Basedow's Disease and Graves' Disease).—A disease of unknown origin,
the cardinal symptoms of which are exophthalmus, enlargement of the thyroid, tachycardia, functional disturbance of

the nervous and vascular systems. Several post-operative cures have recently been reported.

Ferratin.—A red-brown powder, odorless, tasteless, insoluble in water, but soluble in the presence of soda. It contains seven to eight per cent of iron. It is prepared by exposing to a certain degree of heat in an alkaline medium definite proportions of the white of egg and an iron salt. It is a very efficient and prompt chalybeate, easily assimilated, readily taken, and well borne by children and adults. It may be given in powder, capsules, or in the form of chocolate tablets (half gram) as they are sold in the market. The average dose is gr. vii. to viii. three times a day. Now advertised in the public press.

Ferripyrin.—A preparation containing iron twelve per cent and antipyrin sixty-four per cent. Hæmostatic and astringent in twenty-per-cent solutions. Internally given in doses of gr. v. to viii.

Ferrol.—Said to contain:

    ℞ Olei morrhuæ (opt.) .... ℥ vij.
      Ferri phosphat. .... ℥ ij.
      Phosphori. .... gr. i.
      Glycerini. ....q.s. ad ℥ xv.
    M.

Ferrosol.—A black-colored fluid composed of ferric saccharate and saccharate of sodium chloride.

Dose, teaspoonful three times daily in chlorosis, anæmia, etc.

Ferro-Somatose (Iron Somatose).—A combination of iron and albumoses. It consists of an odorless, tasteless powder, soluble in water and aqueous fluids. Being composed of the nutritious elements of meat and two per cent of iron, it acts as a tonic and reconstructive in anæmia, chlorosis, debility, convalescence.

The dose is two teaspoonfuls to a tablespoonful daily for an adult and in proportion for children.

Filmogen.—A new form of varnish said to be superior to collodion, first introduced by Dr. E. Schiff, of Vienna. It consists of pyroxylin dissolved in acetone, to which is added a small quantity of castor oil to render the film more flexible. It is practically the same as the collodion of the U. S. P., only that acetone is used as the solvent in place of alcohol and ether. It is used as a vehicle for various drugs—chrysarobin, ichthyol, menthol, anthrarobin, etc.— and forms a superficial coating.

Filopowicz's Sign of Typhoid Fever.—A yellow color of the palms and soles, seen in the early days of the disease and lasting until the end, is said by Filopowicz (Centralblatt für die medizinischen Wissenschaften, 1898, No. 11) to be of frequent occurrence in enteric fever.

Flechsig's Opium-Bromide Treatment of Epilepsy. (Introduced by Flechsig in 1893).—Opium, beginning with about a grain daily, given in three doses, and steadily increased until four or five grains daily are administered. After a period of six weeks the opium is discontinued, and bromides in large doses are used—seventy-five to one hundred grains a day for at least two months. The fits generally yield to the first dose of bromide. The treatment is applicable only to chronic intractable cases in which the administration of bromides alone has failed. The contraindications to its use are the status epilepticus, plethora, severe heart disease, and cerebral focal lesions.

Flint Murmur.—A murmur of rumbling quality occasionally heard at the apex of the heart in aortic insufficiency. At times it is presystolic.

Florence Reaction.—This is a test for the detection of semen. An iodine-rich solution of potassium iodide and iodine (potass. iodide 1.65 gm., iodine 2.54 gm., distilled water 20 c.c.) is added to fresh semen or to a watery extract of the stains, and then examined microscopically. Numerous dark-brown crystals of various sizes and forms—rhombic tablets, fine needles—are seen. This reaction does not depend upon the presence of spermatozoa, since the same reaction may be obtained from the seminal fluid in conditions of complete azoöspermia. It has been suggested that the crystals come from spermin, a fact which is not altogether improbable, because the addition of the reagent to a solution of spermin phosphate gives typical Florence crystals. This test is not absolute, only corroborative, because substances other than semen give positive results.

Fluoroscope.—An instrument for exploring certain bodies rendered fluorescent by the x-rays. Thus, if in a darkened chamber a Roentgen ray is thrown upon a screen covered with the platino-cyanide of barium or potassium, the tungstate of calcium, etc., the latter becomes luminous. If now between the Crookes tube and the screen an object is introduced—e.g., the hand—the flesh or the parts more transparent to the x-rays will be seen but faintly, while the bones will throw their shadow.

Forchheimer's Exanthem.—This is an eruption which is seen upon the velum and uvula and is said by the discoverer to be characteristic of German measles. It consists of a macular, rose-red eruption, irregularly arranged, of the size of a large pinhead and slightly elevated above the mucous membrane. This eruption extends to, but not over, the hard palate. The eruption lasts twenty-four hours or a fraction thereof. This exanthem is seen only when there is a suggestion of the cutaneous lesion and not in the pre-eruptive stage.

Formaldehyde-casein (Analogous to Glutol). — A feeble antiseptic used as a powder for tampons, gauze, etc. It may be sprinkled on moist dressings at the time of use.

Formalin (Also known as Formic Aldehyde, Formol, and Formaldehyde).—This is a forty-per-cent solution of formaldehyde gas, and consists of a clear, colorless liquid with a sharp, pungent odor and irritating fumes.

It is a powerful antiseptic and disinfectant. Its advantage as a disinfectant lies in the fact that it does not injure the materials subjected to its influence.

A special apparatus is required in order to regenerate the gas in a dry and efficient state from the forty-per-cent solution or from pastils.

The solution must be evaporated in a closed autoclave from which the gas is liberated at a pressure of about three atmospheres, with a corresponding temperature of about 250° F. The presence of a ten-per-cent solution of calcium chloride facilitates the operation.

In disinfecting dwellings, this gas is liberated in great volume, and when the doors are opened it may be impos-

5

sible to enter. The introduction of ammonia gas, however, neutralizes the formaldehyde gas, forming what is known as "formamide."

It has been used in dermatology for tinea tonsurans and circinata, eczema marginatum, mycotic eczema so called, alopecia areata, bites of insects and vermin, hyperidrosis, and bromidrosis. For this purpose it is best diluted in the proportion of one to four, being too irritating in pure form.

For "sweating feet" it has proven very efficacious—far superior, in fact, to chromic acid, powder of salicylic acid, alum, etc. It may be applied in the proportion of one to four, or half a teaspoonful to the quart of water.—GERDECK, *Deutsche mil.-ärzt. Zeitschrift.*

In the proportion of ℔ iv. to ℥ viii. it has been used in atrophic rhinitis. In forty-per-cent strength it is useful for the fixation of blood specimens and the preservation of wine, beer, and pathological specimens.

Fränkel-Gabbet Stain.—A rapid stain for tubercle bacilli. Two solutions are used:

A.

Fuchsin	1
Alcohol	10
Carbolic acid	5
Distilled water	100

B.

Methylene blue	2
Sulphuric acid	25
Distilled water	100

The dried and fixed preparation remains for ten minutes in solution A, is washed in water, dried with filter paper, and is then placed for five minutes in solution B, washed and dried. If the stain has been successful, the preparation looks a faint blue. This differs from Ziehl's stain, in that solution B decolorizes and stains at the same time.

Frémissement Cataire.—Cardiac thrill; most common in mitral stenosis.

Friedreich's Disease (Also known as Hereditary or Juvenile Ataxia).—A disease which occurs in several members of the same family and manifests itself as a rule in early life, usually between the ages of seven and twenty. The principal symptoms are an ataxia in which the entire body—arms, legs, trunk—participates, nystagmus, a peculiar, indistinct, slow, scanning speech, and an absence of the patellar reflex. The pupils react to light and accommodation, there are no motor or sensory disturbances, and the functions of the bladder and rectum are normal.

The disease must be differentiated from tabes and multiple sclerosis.

Frigotherapy.—Thermic agents have in many instances of late been replaced by cold applications with manifest benefit.

The Leiter coil, spraying with chloride of methyl, ether, the Brand bath, etc., are here included. Useful in localized pruritus, urticaria, psoriasis, etc. Inferior to thermotherapy in bactericidal action.

Gaduol.—A substitute for cod-liver oil, possessing the advantage over the latter in not deranging digestion.

Gaertner's Milk (Fettmilch).—A substitute for mother's milk. It looks like ordinary milk with a creamy layer; the reaction is slightly acid, the specific gravity 1.016. It is agreeable to the majority of children and coagulates in finer flocculi than cow's milk. It may constitute the sole diet of the child or form an addition to the breast milk. According to Jacobi it is to be given in the same percentage during the feeding months; the quantities are, however, increased while the number of feedings are diminished. It is occasionally useful in chronic dyspepsia and constipation. It is prepared as follows: A mixture of equal parts of milk and sterilized water at a temperature of 30° to 36° C. is placed into a drum of a centrifuge machine, the Pfannhauser balance-centrifuge if possible. The drum is revolved four thousand times a minute, driving the fat to the centre of the drum because it is in suspension; the sugar, salts, and casein being in solution are uniformly distributed. Two tubes carrying equal amounts of fluid enter the drum, one near the centre, the other at the periphery. By arranging the inner tube at varying distances from the centre, milk containing different amounts of fat may be obtained.

Gaiacyl.—A calcium salt of guaiacol-sulphonic acid. A grayish-violet powder, soluble in water. Local anæsthetic in five- to ten-per-cent solution.

Gallacetophenone.—A yellowish powder derived from pyrogallol. Produces rapid effects in psoriasis in ten-per-cent ointment. It does not stain the linen.

Gallanol.—A product of heating gallic acid with aniline and treating with water made acid with hydrochloric acid. Used in weak solutions (1 : 1,000). It prevents pathogenic power of micro-organisms. Used in diseases of the scalp, in chronic eczema, calming the pruritus, and is said to be superior to chrysophanic acid in psoriasis. It does not stain the linen. Also recommended for favus, skin mycoses, triphophytoses, prurigo. May be used in ointment form or with collodion, traumaticin, filmogen, etc.

Gallol.—Same as gallanol.

Gangolphe's Sign.—A sero-sanguineous effusion in the abdomen in intestinal obstruction, soon after strangulation has occurred.

Garcia Rigo's Process.—A rapid double-staining for blood examination. A drop of blood on a cover-glass is diluted with a drop of simple bouillon (kept sterile by a little formol), the two being stirred with a sterile platinum wire till mixed. The cover-glass is then rested on the end of a slide and carefully warmed over an alcohol-flame for less than a minute. Eosin-stain is next used and washed with water; then methylene-blue and washed again. The specimen is then dried and mounted in Canada balsam, the whole process being accomplished in five minutes under favorable circumstances.

Garofen.—A new vegetable analgesic and antipyretic remedy to replace morphine, acetanilid, etc., said to contain no opium in any form and no coal-tar products.

Gastrectasia.—An enlargement of the stomach associated with abnormal motility.

Gastrodiaphanie.—Intragastric illumination showing by transparence the form, situation, and extent of the stomach. For this purpose a diminutive electric light is fastened in the extremity of a stomach tube, which is introduced into the organ previously filled with fluid by means of a siphon.

Gastroptosis.—A sinking of the stomach, which may be partial or complete: partial, if the lower border alone reaches below the normal; complete, if both upper and lower borders sink. This condition is frequently associated with enteroptosis.

Gelanthum.—A varnish introduced by Unna in 1896. Crude tragacanth is emulsified in the cold for four weeks with twenty times its volume of water. This is then treated with steam for one day and finally pressed through a muslin sieve. Gelatin is swelled up cold and then filtered in a steam filter after long exposure to steam pressure. The mixture of the two together is allowed to swell for two days in steam and then squeezed again through muslin. It is now mixed with five per cent of glycerin, rose-water, and thymol (2 to 10,000).

Advantages.—Ease of spreading; dries more rapidly and with smoother surface than the older skin varnishes; is more cooling, and permits of better suspension and more even distribution of remedies; keeps well if protected from drying.

Uses.—Extensive eczemas, etc.

Gelosin.—A jelly-like substance with which remedies may be incorporated, having great absorbing qualities.

Geographical Tongue.—Also termed wandering rash or exfoliatio areata linguæ. This is a peculiar map-like appearance of the upper surface of the tongue which is frequently seen in young children suffering from the exanthematic fevers, and almost any of the skin diseases—eczema, impetigo contagiosa, miliaria, lupus, tinea tonsurans, tinea circinata, syphilis. It may also be observed in tonsillitis and gastro-intestinal disturbances. Occasionally it is present in several children of the same family. In differential diagnosis it has no corroborative value, nor can one modify the prognosis by its occurrence.

Geosote.—See Guaiacol valerianate.

Among the derivates of Creosote the first place is taken by the Valerianates of Creosote and Guaiacol, Eosote and Geosote.—RIECK.

Gerlier's Disease (Paralyzing Vertigo).—A nervous affection associated with sudden pains in the head and neck, paresis of the extremities, vertigo, ptosis, and depression. Consciousness is retained. It occurs only in men and is epidemic in the canton of Geneva, Switzerland. This distinguishes it from auditory vertigo or Ménière's disease.

Giuffrida-Rugieri's Stigma of Degeneration.—Absent or badly defined glenoid fossa. This condition was found more common in women than in men.

Glandulene.—The name given the dried bronchial glands of sheep, which have been prescribed in tuberculosis and pneumonia, with very contradictory reports. In tablets of gr. iv. each. Dose, three to five tablets three times daily. Each tabloid is said to contain 0.25 gm. of fresh gland substances.

Glénard's Disease.—Enteroptosis.

Gluzinske's Test.—A test for bile pigments. Boiling the solution for a few minutes with formalin gives an emerald-green color, which turns to an amethyst-violet on the addition of a few drops of hydrochloric acid. The author states that this reaction is given by the bile pigments and not by the bile salts.

Glycerophosphate of Calcium.—A preparation which of late has found favor with physicians for the treatment of neurasthenia and hysteria. The best preparation is a French one (Chapoteaut), which may be given in the form of a wine, syrup, or in capsules. The dose is gr. iv. or v., a tablespoonful of the wine or syrup containing gr. iv. In general debility, anæmia, convalescence, phosphaturia, albuminuria, sciatica, tic douloureux, this drug has proven of value. It may be given subcutaneously in doses of gr. iii. to iv. The glycerophosphates of iron, quinine, soda, magnesia, potash, and lithia may also be used in the same conditions.

Glycozone.—A compound of one volume of anhydrous glycerin exposed to the action of fifteen volumes of ozone at ordinary atmospheric pressure. Useful in dyspepsia and gastritis. Given in tablespoonful doses with water. Also used locally much the same as peroxide of hydrogen.

Glutol (Formalin-Gelatin).—Used to form a coating over fresh wounds as a protective. Prepared by exposing gelatin dissolved in water to formaldehyde vapors.

Gmelin's Test.—This is the test for bile in the urine or gastric contents. Put in a test tube or a shallow dish a few cubic centimetres of concentrated nitric acid and add gently a few drops of the bile-containing fluid. The result is a play of colors beginning with green, and followed by blue, violet, red, brown, or yellow. The test is an oxidation one, and the resulting colors depend upon the formation of biliverdin, bilicyanin, and biliprasin, etc.

Golonboff's Signs of Chlorosis.—A sharp pain situated directly over the site of the spleen, and pain on percussion over the ends of the long bones, especially the tibiæ, are said by Golonboff to be characteristic of chlorosis. Splenalgia and osteomyelalgia have also been applied to these symptoms.

Gonagra.—Gout of the knee joint.

Gonangiectomy.—A term used by George W. Johnson to indicate excision of a portion of the vas deferens.

Gonococcia.—A generalized infection of the entire organism with the gonococcus itself or its toxins.

Gonohæmia.—Same as gonococcia.

Gonotoxins.—The toxins of the gonococcus of Neisser. These may act upon the blood, kidneys, nervous system, and skin. Experiments go to show that injections of these toxins do not produce immunity to gonococcic infection.

Gousset's Symptoms of Phrenic Neuralgia. — The existence of a painful point, which is always present and well marked to the right of the fourth or fifth chondrosternal articulation. This must not be confounded with the retrosternal pain of chronic aortitis.

Gowers' Hæmoglobinometer.—An apparatus for determining the hæmoglobin percentage in the blood. It consists of (1) a graduated pipette (*mélangeur*) with india-rubber tubing and mouthpiece; (2) of a small graduated

tube with markings ranging from 10 to 140; (3) two small
tubes containing a solution of picrocarmin glycerin, one
with a blue tip for working at night, the other with a yellow tip for day work: (4) a small pipette for water. In
making the test, the finger is cleansed and pricked with a
needle; blood is sucked up to the mark in (1) and then
blown into (2); the blood is then diluted gradually with
water by means of (4), and when its color corresponds to
the picrocarmin-glycerin solution in (3) we read off the
percentage of hæmoglobin. It is advisable to place a few
drops of water in (2) before beginning the test. The tube
containing the blood and the one with the test solution are
best examined against a white background, otherwise errors
of refraction may give a wrong result. This apparatus is
inexpensive ($2.50 to $3) and while not quite so accurate
perhaps as Fleischl's instrument, is for all intents and
purposes just as good for practical and ready examinations. It might be added that a perforated piece of rubber
for holding the tubes goes with each outfit.

Graefe's Sign of Basedow's Disease.—Lack of normal relation between the movements of the eyeball and
upper lid. When the eyeball is moved downward the
upper lid fails to follow it. This sign may be the earliest
symptom of the disease.

Graphorrhœa.—A flow of written words. Analogous
to logorrhœa. Observed as an intercurrent or intermittent
condition in some forms of insanity in which the subject
is seized with uncontrollable desire to cover page after
page with strings of words usually unconnected and mean-
ingless.

Graphospasm.—Another term for writer's cramp.

Guacamphol.—The camphor ester of guaiacol used in severe night sweats of phthisis and diarrhœa.

Guæthol.—An analgetic oily liquid, said to be superior to guaiacol. Used in ointment, one part to six of vaseline.

Guaiacetin.—The soda salt of pyrocatechin mono-acetate. It is a white odorless powder, with a bitter taste, and soluble in water and wine. It has been used in doses of gr. v. to x. for the treatment of phthisis. It may be administered in powder, capsule, or tablet form, or in a mixture containing sherry.

Guaiacol.—A fluid having the odor of creosote, obtained from guaiacum resin. Useful in phthisis. Dose, ℥ ss. to ii. several times a day after meals. Externally, mixed with glycerin or vaseline to relieve the pain and swelling in orchi-epididymitis, acute articular rheumatism, and other local inflammatory processes. Applied with camel's-hair brush cautiously.

Guaiacol Carbonate.—A white, crystalline powder which is odorless, tasteless, insoluble in water. Used with success in phthisis, increasing the appetite, reducing the fever, and diminishing the cough. Usually given in capsules, gr. v. three times a day. The dose may be increased to ℨ i. daily. It is said not to be decomposed until it reaches the intestine.

Guaiacol Phosphate.—A colorless, odorless, crystalline body, given in doses of gr. vii. to x., especially in pulmonary tuberculosis. Best administered in cachets, pills, or capsules.

Guaiacol Valerianate.—An oily liquid used in pulmonary affections and for subcutaneous injection in tuberculous joints and nodes.

Guaiaquin.—A proprietary preparation, being the guaiacol bisulphonate of quinine and said to possess the double properties of guaiacol and quinine. It is a yellowish, bitter powder, soluble in water, alcohol, and dilute acids. It has been used in malaria, typhoid fever, anæmia, cachexia, and sepsis. Dose, gr. v. to x. three times a day. Best given in pill form with gelatin coating.

Güaranteed Milk.—Milk furnished by dairymen, farmers, and milk laboratories with a guarantee that the cows from which it is obtained have been tested and failed to react to tuberculin, and that it is otherwise pure.

Gull's Disease.—Generally known as myxœdema.

Gunzburg's Solution.—This solution is used for testing the presence of free hydrochloric acid in the stomach contents. It consists of 2 gm. of phloroglucin, 1 gm. of vanillin, in 30 gm. of absolute alcohol. To make the test, a few drops of the filtrate of the stomach contents are placed in an evaporating dish or a china spoon, and two or three drops of this solution are added; the mixture is heated over a spirit lamp or Bunsen burner, when, if free hydrochloric acid is present, streaks of red will appear along the side and on the surface of the mixture. The test is a very delicate one, reacting with less than one part of hydrochloric acid in 10,000 of water. In cases in which there is but a slight amount of free hydrochloric acid it may be necessary to blow upon the spoon or dish in order to bring out the red margins. The fluid is to be kept in a dark place, and only a few drops need be used for any one test.

Gynecomastia.—Excessive development of the male breasts.

Gyromele.—An instrument devised by Turck of Chicago. It is a revolving sound with a sponge attached to the cable, worked by a wheel like that of an egg-beater or surgical drill. Before being introduced into the stomach the cable is slipped through a rubber tube long enough to reach through the œsophagus to the cardiac opening. When in operation the rotating and vibrating cable may be felt by the hand upon the abdomen to pass along the greater curvature and then the lesser curvature, the cable adapting itself to the shape of the stomach. Besides its use in diagnosis it is said to indicate the character of the stomach wall, to show the presence of atrophy, and to be useful in therapeutic ways.

Hæmato-Colpos, Metra, Salpinx.—Conditions observed in atresia of the vagina or hymen, and consisting respectively of the distention of the vagina, uterus, and tubes with blood.

Hæmatoporphyrin.—The following test for hæmatoporphyrin in urine has been given by A. Garrod: To 100 c.c. of urine add 20 c.c. of a ten-per-cent sodium-hydrate solution, and filter. Wash the filtrate thoroughly with water, add absolute alcohol, then enough hydrochloric acid completely to dissolve the precipitate. Examine the solution with the spectroscope for the (two) characteristic hæmato- porphyrin absorption bands.

Hæmochromogen.—Obtained by reducing with hydrazin hydrate a solution of hæmatin in a moderate amount of ammonia. This is washed with alcohol and ether, and then dried to a fixed weight, giving a brownish-red powder with fairly stable properties.

Hæmol.—Blood of cold-blooded animals neutralized with water and powdered zinc is precipitated with sulphate of ammonium, and hæmol is thrown down by the aid of hydrochloric acid. It is a brown powder, easily taken and rapidly assimilated, and said to be quickly transformed into the coloring matter of the blood. Dose, gr. ii. to viii., three times a day in wafers or capsules. Useful in debility, anæmia, chlorosis, and intestinal ulcerations.

Hæmostaticum.—A thymus extract containing seven per cent of calcium chloride, made alkaline with liquor sodii.

Haeser's Coefficient.—See Haeser's Rule.

Haeser's Rule for estimating the solids in the urine consists of multiplying the last two figures of the specific gravity by 2.3 (this is known as Haeser's coefficient), and then multiplying again by the number of litres passed in twenty-four hours. The answer represents the amount, in grams, of the solids passed in twenty-four hours.

Haffkine Virus.—An antidote for cholera by inoculation. The true worth of this remedy has not become sufficiently known from tests so far made.

Hager-Schmidt Treatment of Tænia.—

R Oxyd. nigr. cupri.. 6
  Cretæ prep... 2
  Argillæ alb... 12
  Glycerini 10
M. ft. pil. No. cxx.   S. Two pills four times daily.

Haine's Modification of Haeser's Rule.—Multiply the last two figures of the number representing the specific gravity by the number of ounces passed in twenty-four hours, and to this add ten per cent of the result. The product represents the number of grains of solids voided in the twenty-four hours.

Hammer-Toe.—Hyperextension of the first phalanx upon the metatarsal bone and forced flexion of the last two phalanges on the first.

Hansen's Bacillus.—The bacillus of leprosy. Stains with Ziehl's solution (see p. 183).

Harris' Instrument.—An instrument devised by Dr. M. I. Harris for obtaining the urine separately from both kidneys. It consists of a double catheter enclosed in a common sheath about 7½ inches long, each catheter being separately movable within the sheath, which is graduated in centimetres. At the end of each catheter a short piece of rubber tubing is connected with separate glass vials, with two corks, one for the tubing to the catheter, the other for tubing to a rubber exhaust bulb.

Harrowing.—A method first used by Dr. Dalagénière for the treatment of sciatica, consisting of tearing apart or teasing the fibres of the nerve with any blunt instrument.

Hayem's Solution.—One of the best fluids for use in counting the red blood cells:

Bichloride of mercury....	0.5
Chloride of sodium	1.0
Sulphate of soda	5.0
Distilled water	200.0

The fluid is used with the Thoma-Zeiss counting apparatus in the same manner as chloride of sodium.

Hayem's Solution for intravenous or subcutaneous infusion:

Sodium sulphate...............................   10
Sodium chloride.............................   5
Distilled water 1,000

Used in severe hemorrhages, shock, uræmia, eclampsia, and toxæmia.

Hazelin.—A preparation of witch hazel (Hamamelis virginica). It is anodyne, astringent, and styptic.

Heckel's Prasoide Tincture.—Recommended by Balestre in gout. It contains globularin and globularetin in proportion of 15 to 17 cgm. per teaspoonful. Dose, one to four teaspoonfuls daily. As preventive six drops morning and night, increasing gradually to twenty or twenty-five at a dose. To be continued for months with occasional interruptions.

Hegar's Method of Diagnosticating Fibroma.—The finger is passed into the rectum and pressed against the tumor; at the same time the uterus is drawn downward by a volsellum. If the tumor is ovarian, it will not move; if uterine, there will be great resistance to drawing down the cervix, which will clearly be continuous with the morbid growth.

Hegar's Sign.—Softening of the lower isthmus of the uterus in the later months of pregnancy.

Heintz's Method for the Determination of Uric Acid.—To 200 c.c. of urine add 10 c.c. of hydrochloric acid. Let it stand in a cool room twenty-four hours. Collect the precipitated uric-acid crystals on a weighed filter, and wash with cold distilled water. Dry filter and crystals in a desiccator and weigh. Subtract the weight of filter, and the result will be the weight of uric acid in 200 c.c. of the urine. If albumin is present it must be removed, and filtration must always be done before applying the test, otherwise subsequent filtration will be very difficult.

Heliotherapy.—The application of burning-glasses to cancer, tuberculous ulceration, or to the destruction of microbes and the results of their action. It is essentially the transmission of solar light and heat in a concentrated form by means of lenses. The method was known to Pliny, who speaks of using the sun's rays as a caustic by means of crystal globes.

Heller's Test for (*a*) albumin and (*b*) blood in the urine:

(*a*) To 1 c.c. of nitric acid in a test tube add by means of a pipette a few drops of urine; if albumin is present a white ring will appear at the junction of the two fluids. The intensity of the ring is directly proportional to the amount of albumin present.

(*b*) To a few cubic centimetres of urine add a solution of liquor potassæ, and heat gently. The phosphates which become precipitated carry the blood down with them and settle at the bottom as a reddish mass.

Hemamœba Leukæmia.—A term which Loewit thinks
6

future investigation may bring into use to designate the association of an organism belonging to the protozoa (probably acystosporidei) in the blood of leukæmic patients. It seems to subsist upon the leucocytes, but may live in the plasma. It is generally amœboid, and has been demonstrated in the spleen during life by puncturing the organ.

Hemianopsia.—Blindness in half of each eye. If the blindness be in like-named halves of both visual fields it is termed homonymous hemianopsia; if in opposite, heteronymous hemianopsia.

Hemicranin.—A proprietary preparation, said to be a mixture of phenacetin, caffeine, and citric acid in the proportion of 5, 1, and 1. Used in gr. x.-xv. doses as an analgesic and sedative in migraine and neuralgia. The maximum daily dose is ℨ iss.

Henoch's Disease (Nervous Purpura).—This is a disease of childhood, and begins suddenly with severe colicky pains, diarrhœa, and vomiting; this is followed by swelling of the knees, ankles, shoulders, elbows, and wrists, which is accompanied by pain and petechiæ on the extensor surfaces. Internal hemorrhages may occur.

Hepatic Extract.—Gilbert and Carnot have shown the coagulating action of liver extract on the blood, and their observations have been corroborated by Berthe (*Jour. de Méd.*, October 10, 1898). The latter used it in hæmoptysis, epistaxis, and metrorrhagia with gratifying results. The dose is ℨ iii., given in tepid soup, repeated if necessary. It may also be administered per rectum. It has been given in the form of an alcoholic, aqueous, or glycerin extract, but the best preparation is desiccated liver. Fresh pig's liver is one of the best sources for procuring the preparation.

Hepatopexia.—Fixation of a movable liver.

Heroin.—A di-acetic derivative of morphine by substituting groups of acetyl for two hydroxyl groups. A substitute for morphine and codeine. Sedative action on the respiration greater than that of morphine. It is said to be ten times more powerful than codeine and only one-tenth as poisonous. Efficacious in coughs in dose of gr. $\frac{1}{12}$ to $\frac{1}{6}$. In affections marked by shortness of breath it is said to cause deeper and easier respiration. Floret states that it allays cough and pain in angina, bronchitis, and pharyngitis. Manges (New York *Medical Journal*, November 26, 1898) has found it very efficacious in bronchial asthma, pleurisy with effusion, incipient tuberculosis, emphysema, and pneumonia. Occasionally mild disturbances of the sensorium, especially in old people, followed its use. It also has a mild antipyretic action. Being insoluble in water the drug is best administered in the form of powder, capsule, or tablet triturate. Dose, gr. $\frac{1}{20}$ to $\frac{1}{6}$. Average dose, gr. $\frac{1}{12}$.

Hersage.—A dilaceration of the sciatic nerve for a rebellious sciatica.

Hetol (Cinnamate of Sodium).

Hippus.—An alternate contraction and dilatation of the pupil due to a clonic spasm of the iris and seen occasionally in hysteria, multiple sclerosis, and migraine.

Hirsh-Libman Streptococcus.—Streptococcus found in severe diarrhœas of infancy and having the following characters: "In the fæces it appears as a flattened coccus, in short straight chains or angular chains, the diplococcus arrangement being frequent. In sugar bouillon, after twenty-four hours there is a uniform cloudiness; after five or six days the upper part of the fluid clears. The growth is less active in ordinary bouillon. In bouillon the coccus appears frequently in groups of two or three. In longer chains, one of the chains is apt to be larger. The average size of the cocci, stained in fuchsin, is 0.75 μ by 0.9 μ. The coccus stains with Gram solution."

Hiss Method.—A method by means of which it is possible to obtain within forty-eight hours the typhoid bacilli from the fæces and urine of a patient suffering from the disease. Two media are used, one by plate culture, the other by tube culture. On these media—composed of agar, gelatin, sodium chloride, meat extract, and glucose slightly acidulated with hydrochloric acid—small colonies form with irregular outgrowths and fringing threads. The colon colonies on the other hand are larger, darker, and, as a rule, without threads.

Holocaine.—A substitute for cocaine in ophthalmology and laryngology. It is an alkaloid obtained synthetically from paraphenetidin. It is soluble in water and used in one- to four-per-cent solutions. It is relatively cheap and non-toxic, does not affect the accommodation or contract the blood-vessels. In the throat it does not produce a bitter and nauseating sensation, and does not reduce swelling.

Holzin.—A sixty-per-cent solution of formaldehyde in methyl alcohol; first used by Opperman in the treatment of whooping-cough.

Homatropine. — White crystalline flakes soluble in water. Used almost exclusively in eye practice. The solution employed is from two to six grains to the ounce. A few drops of this produces dilatation of the pupil. Its advantages over atropine are the greater rapidity of dilatation, the earlier disappearance of its effects, and the lesser danger of increasing the eyeball tension.

Hot-Air Treatment. — Also known as Tallerman's Method.

Huntingdon's Chorea.—A chronic form of chorea occurring late in life and characterized by irregular movements, disturbance of speech, and psychical defects. It appears, as a rule, between the twenty-fifth and fortieth year, and is occasionally hereditary.

Husa (Also called *yousa*, *hoosa*, and *yusce*).—"This is an unclassified plant of a dirty whitish-green color, about two or three inches long. It is possibly indigenous to the Everglades of Florida."—WINTHROP.
This plant has been said to act as a remedy for the opium habit and as an antidote to snake-bite, insect stings, etc.

Hydrargyrum Colloidale.—A soluble mercurial introduced by Lottermoser, investigated by Werler (*Berlin. klin. Woch.*, October 17, 1898).

Hydroleine.—A preparation of oleum morrhuæ. Each dose of ℥ ii. is said to contain:—

Pure Norwegian cod-liver oil................ ℥ lxxx. (gtt.)
Distilled water .... ℥ xxxv.
Soluble pancreatin .... gr. v.
Soda...................................... gr. $\frac{1}{3}$
Salicylic acid....... .... gr. $\frac{1}{4}$

Hydrophthalmia.—The glaucoma of children.

Hyoscine.—The hydrobromate is used hypodermatic-ally in doses from $\frac{1}{200}$ to $\frac{1}{100}$ in acute mania, chronic de-mentia, delirium tremens, and paralysis agitans.

Hypnal (Hypnol).—A crystalline, odorless, tasteless substance produced by mixing chloral and antipyrin. It soothes pain, relieves coughing, and produces sleep. It is a good and safe hypnotic and does not irritate the stom-ach. Dose, gr. xv. to xxx.

Hypo-Quinidol.—A combination of quinine and phos-phorus. Dose, same as that of quinine, in tonic dose.

Ichthalbin.—A mixture of ichthyol and albumen em-ployed for internal use in gastro-intestinal disorders. Dose, gr. v. to viii.

Ichthyol (Sulpho-ichthyolate of Sodium or Ammonium). —This is a product of a bituminous mineral containing the residues of fish and other substances found at Seefeld in the Tyrol. The commercial article is obtained by distilla-tion, first treating with sulphuric acid, then neutralizing with soda or ammonia. It contains an equivalent of ten per cent. of sulphur. It is a tarry liquid of a strong odor.
This drug is an antiseptic, deoxidizer, and probably the best application for erysipelas; it is used in many skin

affections and internally, preferably in pill form, in uterine and intestinal affections. Internal dose, gr. ii. to v.

Ictus Laryngé.—This term is applied to an apoplectiform attack which occurs during a severe paroxysm of coughing, and which after a few seconds passes off without leaving the slightest trace of unconsciousness. In some instances small doses of antipyrin over a long period have given good results.

Idioglossia.—A disorder of speech in which the impression is made that the speaker is employing a language of his own. The term lalling has also been suggested for the same tendency to mispronounce vowel sounds and to substitute one consonant for another.

Indican.—Also known as indoxyl-sulphate of potassium. The precursor of this substance is indol which is normally formed in the intestine by the action of bacteria upon proteids. The latter is oxidized in the tissues into indoxyl, which unites with sulphuric acid, forming indican or indoxyl-sulphate of potassium. In health indican is found in but small quantities in the urine, so that its presence in large amounts is to be looked upon as pathological. Most generally its appearance points to an increased decomposition in the intestines; thus, in habitual and prolonged constipation, typhoid fever, cholera infantum, carcinoma of the caput coli, and intestinal obstruction large amounts are found. In fact, wherever decomposition of albuminous substances takes place—putrid bronchitis, gangrene of the lung, empyema, suppurative peritonitis—indican occurs in large quantities. As a rule, the urine is colorless when passed; in some cases, however, it looks blue.

Tests.—The two tests generally employed are those of Jaffé and Obermayer. Both are oxidation tests and have for their object the splitting up of indoxyl-sulphate of potassium and the setting free of a colored product—indigo blue.

Jaffé Test.—To 5 c.c. of urine add an equal quantity of hydrochloric acid and one drop of a ten-per-cent solution of hypochlorite of soda or calcium hypochlorite; now add an excess of chloroform and shake well. If indican is present the solution assumes a blue color, because the indigo blue has been set free and has been dissolved in the chloroform. An excess of the hypochlorite solution must be avoided, for otherwise hyperoxidation will take place and the solution appear colorless even though indican is present.

Obermayer Test.—To the urine add an excess of a twenty-per-cent solution of plumbi acetas, and filter; now add equal part of Obermayer's reagent (5 gm. sesquichloride of iron, 500 gm. fuming hydrochloric acid), and then chloroform. Shake well and the blue color appears. The advantage of this test is that there is no danger of hyperoxidation.

Intracranial Injections of tetanus antitoxin have been employed with success in severe lockjaw. The first use of the method in this country was at the general hospital in Passaic, N. J., in October, 1898.

Invalid Bed.—A combination bed and reclining-chair, ·the latter turning on a pivot so as to stand at right angles to the bed and permit the patient to sit with the feet projecting over the side of the bed, has been devised by Ernst Leutz (see cut, *Zeitschr. für Krankenpflege*, August, 1898).

Iodine, Test for.—In some instances, such as the appearance of high fever, collapse, delirium, vomiting, etc., after the application of iodoform over large extents of surface, in the form of gauze, powder, or ointment, it becomes necessary to test for iodoform in the urine. For this purpose (*a*) a few drops (three or four) of nitric acid are added to the urine in a test tube, and this is followed by half a teaspoonful of chloroform. The test tube is then well shaken and the fluid assumes a red-violet color, with separation of crystals.

(*b*) A small pinch of powdered calomel is placed upon a white saucer, and then a few drops of the urine to be examined are dropped upon it; a mixture of the urine and calomel is then made with a glass rod. If the urine contains a notable amount of iodine there is produced a well-marked yellow discoloration, which should indicate that the iodoform is being absorbed in sufficient quantity to produce danger.

Iodofats.—Substances obtained by Winternitz, by treating various kinds of fats with iodine monochloride. These fats contain equal parts of iodine and chlorine. After feeding with these fats iodine is found in the urine in the form of an alkaline iodide and also in organic combination.

Iodoformin.—A red crystalline powder containing eighty per cent of iodine.

Iodoformogen.—Said to be a combination of iodoform and albumen. It is a bright yellow powder, odorless, soluble in water, and used as a dusting-powder for wounds. It may produce a dermatitis.

Iodol.—An odorless, tasteless, light brown or yellow

powder, occasionally used as a substitute for iodoform.
It contains less iodine than the latter, and is far less poisonous.   It is almost insoluble in water, but soluble in six
parts of absolute alcohol.   Used in chronic conjunctivitis,
ulcus molle, simple ulcers, etc.   In nasal work one per
cent of menthol added is said to enhance its action and to
improve the odor.

Iodol-Menthol.—Said to be a mixture of iodol with
menthol (one per cent).   Applied in the form of insuffla-
tions for nose and throat work.

Iodospongin.—A proteid substance containing iodine
and isolated by Harnack from the ordinary bath sponge.

Iodothyrin.—A combination of the active principle of
thyroid gland and sugar-of-milk.   It consists of a whitish
powder, with a sugar-of-milk taste.   It has been recommended in myxœdema, cretinism, strumous diathesis,
obesity, uterine hemorrhages, psoriasis, rickets, etc.   Dose,
gr. v. to x. three times daily for adults; for children, gr.
x. to xv. a day.   It is excreted in the milk.

Ivan Bang, of Christiania, has reported a case of a child
with congenital struma in whom the neck circumference
was diminished to a great extent during the period of
nursing; the mother had an enlarged thyroid, and was
being treated with iodothyrin.

Iodozone.—A solution of iodine in ozone.   Used as a
spray for wounds, and has been recommended in pulmonary tuberculosis.

Iophobia.—Morbid fear of poisons and virus.

Ischæmia.—Local anæmia.

Ischæmia, Artificial, produced by mechanical means, as by position, application of Esmarch bandage, etc.

Izal.—A coal-tar-derivative antiseptic of recognized power. Ointments, surgical dressings, soaps, and recently "perles" for internal use have been placed on the market, especially in England. It resembles creolin.

Jacob's Ulcer (Rodent Ulcer).—This is an epithelioma of a non-malignant, local character, which begins as a pimple or wart on the skin near the inner canthus, and finally develops into a shallow ulcer with a well-marked, sclerotic margin. It is rarely seen before the fortieth year.

Janet Method.—A plan of irrigation of the urethra with permanganate of potassium of varying strength (1:5,000 to 1:1,000). A progressive step in the treatment of gonorrhœa which some regard as at times abortive. Janet's special apparatus is used. By closing the meatus the fluid flows back into the bladder.

Jelloids.—A form of coated pill in which jujube mass is used as the protecting cover.

Jorisenne's Sign of Pregnancy.—An acceleration of the pulse beat on assuming the erect posture after rest. It is of doubtful value.

Kaiserling's Method for the preservation of specimens with their natural colors. The organ to be preserved is cut into slices from 3 to 5 cm. thick which are placed for from three to five days in—

1.

Formalin	200 c.c.
Water	1,000 c.c.
Nitrate of potassium	15 gm.
Acetate of potassium	30 gm.

They are then removed, the fluid is allowed to drain off, and are placed in—

2.

Alcohol, 80 per cent, for six hours; then
Alcohol, 95 per cent, for two hours.

From this directly into—

3.

Water	2,000
Acetate of potassium	200
Glycerin	400

for permanent preservation in a dark place. Further details for the preservation of whole organs, etc., should be studied in the original article (C. Kaiserling, *Virch. Arch.*, Bd. 147, S. 389). The results thus obtained are certainly wonderful and promise to be one of the greatest of this generation's aids to teaching (*Journal of the Boston Society of Medical Sciences*, March, 1897).

Kakke.—The Japanese term for beri-beri.

Kala Azar.—An epidemic malarial fever peculiar to the region of Assam.

Kaolin.—Kaolin preparation (bolus alba of the German and Japanese pharmacopœias). Native white silicate of alumina. Useful absorbent powder in dermatology, and may be used to dilute permanganate of potassium, silver salts, etc., in coating pills or powders, since it is not acted upon by chemical reagents.

Kaori.—A paint for the dressing of wounds.

Kaposi's Disease.—Xeroderma pigmentosum.

Katatonia.—First used by Kahlbaum to denote a type of melancholia beginning with vertigo, insomnia, abnormal sensations in the head, irritability, and gradually increasing melancholia.

Keratin.— A substance obtained by treating horn shavings with ether, alcohol, and an acid. It is used for coating pills which are intended for medication of the intestines. This substance is not acted on by the gastric juice, but freely dissolves in the intestinal contents.

Kernig's Phenomenon of Meningitis.—A person affected with meningitis when raised to the sitting posture, cannot extend the slightly flexed leg, whereas in the lying position the limb is perfectly relaxed. Netter (*Progrès Medical*, July 30, 1898) regards this sign as almost pathognomonic, and has observed it forty-one times in a series of forty-six cases. A person in health when sitting up in bed with his leg flexed can straighten it without any difficulty. This sign was absent in typhoid fever, pneumonia, polyarthritis, and chorea. Netter has found it present in all forms of meningitis—tuberculous, secondary, and cerebro-spinal.

Kineto-Therapeutic Bath.— Water gymnastics or baths in which certain motions and movements are carried out. Advocated in paresis, muscular atrophy of the lower extremities, etc.

Kirstein's Autoscope.—An instrument somewhat of the shape of a vaginal retractor, used for direct inspection of the vocal cords, by elevating the epiglottis.

Kola-Tannin.—A compound of caffein and tannin, containing twenty to twenty-five per cent of the former, obtained by Prescott and Knox from kola-nuts.

Kolpitis Emphysematosa (Zweifel).—Emphysema of the vagina, first described by von Winckel, by whom it was called kolpohyperplasia cystica; it is characterized by gas-containing cysts in the vagina, especially in pregnancy. There is a feeling of dryness and crackling on examination. Suggested by J. M. Jackson to be due to blocking of lymphatics, especially as it is seen after abdominal hysterectomy.

Koplik's Spots.—A pre-eruptive and early eruptive pathognomonic sign of measles, which consists in the appearance "on the mucous membrane lining the cheeks and lips (buccal mucous membrane) of small irregular spots of a bright red color, in the centre of which there is a minute bluish-white speck.

"This eruption or phenomenon or sign appears on the mucous membrane of the cheeks and lips fully twenty-four or forty-eight or even seventy-two hours before the appearance of the eruption or exanthema of the skin. It spreads and reaches its height just as the skin eruption has appeared and is spreading. It then fades. Thus we have a sort of a cycle."

The number of spots varies from a dozen or more to the lining of the entire cheek. The spots never coalesce, however, the punctate character always being retained.

"If the mouth has been washed these spots may have been rubbed off and then the appearance is lost." In order to see these minute spots distinctly very strong daylight must be used, and the mucous membrane of the cheeks and lips must be everted with the fingers or a spatula, preferably the latter. Microscopically the spots are said to consist of fatty epithelium. Thus far neither fibrin nor bacteria have been found. The number of these spots bears no relation to the severity of the disease and does not influence the prognosis one way or the other.

These spots are of value, not only in diagnosticating an incipient measles to the exclusion of influenza, bronchitis, tonsillitis, febricula, etc., but also in differentiating morbilliform and other eruptions from measles. In a large series of dermatological cases the authors have never seen this sign in any other disease.

Koumiss (also spelt Koumyss).—A nutrient, stimulating, and effervescent beverage which is essentially fermented mare's milk. It contains alcohol, carbonic-acid gas, sugar, fat, and lactic acid, and may be prepared as follows: To one pint of cool milk add two teaspoonfuls of sugar of milk and then one-sixth of a cake of Fleischmann's yeast. Cork the bottle tightly and allow it to stand in a warm place (70° to 80° F.) for eight to ten hours; it is then to be put on ice and used when cold. It is indicated in phthisis, chlorosis, anæmia after hemorrhage, and in all protracted diseases, malnutrition, and cachectic conditions. In cases of irritable stomach it is often retained after everything else fails. It should be taken cold, and thus administered it often acts as an anti-emetic. It should not be used as a substitute for milk alone, since it contains from two to four per cent of alcohol.

Kreosolid.—A magnesium compound of creosote, used by Denzel in tuberculosis.

Kristalline.—A substitute for collodion prepared by dissolving gun-cotton in methyl alcohol. Like collodion and traumaticin it is used as a vehicle for various drugs applied in skin medication—chrysarobin, pyrogallic acid, aristol, anthrarobin, gallanol, etc.—forming a firm, transparent film.

Kryofin.—This occurs in the form of fine, colorless, odorless, and tasteless crystals, soluble in hot water, alcohol, and glycerin. It is an antipyretic, analgesic, and antineuralgic powder in doses of gr. iv. to viii. Useful in migraine, neuritis, myalgia, meningitis, articular rheumatism, influenza, and the lancinating pains of tabes dorsalis. It may be given in powder, capsule, water, or aromatic syrups. Highly recommended by Eichhorst and Mink.

Kummerfeld's Solution.—

R Sulph. precip .... ℥ i.-iij.
Pulv. camph .... gr. v.
Pulv. tragacanth .... gr. x.
Aquæ calcis,
Aquæ rosæ................................. āā ℥ i.
S. Apply after washing at night.

Used in varying strengths for acne, according to the susceptibility of the skin and the severity of the case.

Laborde's Method of resuscitation in apparent death from drowning and anæsthetics. This method really consists of rhythmical traction of the tongue. Place a piece of linen round the tip of the tongue and grasp it with the thumb and middle finger; now pull the tongue forward with a jerk and then relax it again; repeat this manœuvre twenty times a minute. A sense of resistance is felt in the tongue before there is any attempt at respiration. Traction should be continued for thirty or sixty minutes. Dr. W. Freudenthal (*New York Medical Journal*, December 10, 1898) refers to his modification of this method, which consists of irritation of the epiglottis with a view of producing reflex action. He moves the finger to and fro over the epiglottis, thus "tickling" it, as it were.

Lactic Acid.—Used by Zolatorin with good results in a case of arthritis deformans; the dose was forty drops daily for three weeks. Also advocated for laryngeal tuberculosis and leucoplakia buccalis. Bowles has used it for summer diarrhœa in doses of gr. 1¼ every hour.

Lactic-Acid Bacilli.—These are long, thread-like, immobile bacilli, first described by Boas and found most frequently in the gastric contents of carcinoma of the stomach.

Lacto-Glycose.—A dry powder prepared from Mellin's Food and milk, free from starch and with the casein mechanically broken up.

Lactol (Lacto-Naphthol).—Analogous to benzo-naphthol.

Lactone.—An unfermented sparkling milk product.

Lactopeptin.—A grayish powder used as a digestant, in doses of gr. x. to xx. in powder or tablet form, said to contain pepsin, ptyalin, pancreatin, lactic and hydrochloric acids.

7

Lactophenin.—Said to be a safe and efficient antipyretic and sedative, particularly adapted for children. The dose for a child one to two years old is from gr. ss. to i. three times daily. The average adult dose is gr. v. to xv. repeated every three hours. The maximum daily dose is ʒ i. It may be given in powder or tablet form as such, or combined with quinine, phenacetin, codeine, Dover's powder, caffeine, etc. It has been tried and found very efficacious by Caillé, Jaksch, Landowski. G. von Roth (*Wiener klin. Wochenschrift*, vii. 37) has used it in acute articular rheumatism and considers it as efficacious as salicylate of sodium. It occasionally produces a transitory cyanosis.

Lactose.—As a diuretic in doses of ʒ iii. daily, dissolved in two quarts of water.

Lacto-Somatose (Milk Somatose).—A combination of the albuminous principles of milk with five per cent tannic acid. It is an odorless and tasteless powder, soluble in water and aqueous solutions. Dose for children, one to two teaspoonfuls; for adults, two to three tablespoonfuls. Used especially in children for gastro-intestinal disorders.

Lævulose.—See Diabetin.

Lævulosuria.—A variety of diabetes. The elimination of lævulose by the urine seems to give rise to a syndrome in which certain symptoms common to grave neurasthenia predominate (insomnia, the idea of crime, tendency to suicide). If not promptly recognized suicide may occur. In testing be guided by Selievanoff reaction.—ROBINSON.

Lalande's Method for the Treatment of Syphilis.—Dr. Lalande, a homœopathic physician of Lyon, re-

cently proposed before the Société de Biologie de Paris to
treat syphilis with hypodermic injections of a saline solution of powdered calves' horn in the proportions of

Pulv. cornu	60 gm.
Sod. chlor	10 "
Aquæ dest...................................... 1,000	"

The doctor claims the most satisfactory results for his
prescription; old cases in which he used the treatment for
two years had no relapses during that time. Also known
as the keratin treatment.

Landry's Paralysis (also known as Acute Ascending
Paralysis).—This is an affection characterized by the sudden paralysis of the lower, then the upper extremities,
and finally the medulla oblongata, while sensation and the
functions of the bladder and rectum remain intact. It is
a disease of adult life and usually ends fatally, death being
due to respiratory paralysis. The disease generally lasts
from seven to fifteen days. Some cases are only with difficulty distinguished from multiple neuritis and myelitis.

Langier's Sign.—In obstruction of the small intestine
the abdomen is globular in the centre and flat in the flanks.

Lanolin Powder.—A mixture of zinc oxide and magnesium carbonate with lanolin. It is prepared by dissolving the lanolin in ether, adding the powder, evaporating, and then pulverizing the residue.

Lanoform.—Lanolin mixed with one per cent formaldehyde. Used as an antiseptic.

Largin.—An albumen-silver preparation used in the
treatment of gonorrhœa. It is a grayish powder produced
by the action of an ammoniacal solution of silver oxide on

an alcoholic solution of the dry product of decomposition
of the paranucleo proteids.   Pezzoli (*Wien. klin. Wochen.*,
No. 11, 1898) states that as an antigonorrhœic it is equivalent to the other silver preparations, that it kills gono-
cocci more promptly, and that it penetrates more deeply
into dead organic substances.   It is used in a solution of
1 : 6,000 to 1 : 4,000, or stronger.

Lasegne's Sign.—In sciatica, if the thigh is flexed upon
the pelvis with the leg fully extended at the knee, considerable pain is produced, because the sciatic nerve is thus
stretched.

Lehman's Sign.—To prognosticate as to an easy or
difficult anæsthesia in giving chloroform, if the lids closed
by the anæsthetizer reopen at once, wholly or in part, it is
an indication that the narcosis will be difficult.   Con-
versely, in those who take chloroform well the eyes remain
closed from the beginning.

Leinol.—Recommended by Prof. William H. Thomson
for chronic coughs and colds.   Each fluid ounce is said to
contain :

```
Olei lini comp.................................. 33⅓ per cent.
Acidi hydrocyanici........................... gtt. iv.
Codeinæ sulph................................. gr. ss.
Olei cinnamom.,
Chondri....................................āā q.s.
```

Leistikow's Lotion.—

```
℞ Corrosive sublimate ..............................   0.5
   Alcohol,
   Chamomile water ..........................āā  25.0
   Chloroform...................................... gtt. v.
   Cherry-laurel water.......................q.s. ad 100.0
```

This is used for pruritus of the scrotum.

Leiter Coil.—A continuous rubber tube coiled into a circular pad to be applied to the body's surface. By inserting one end in a vessel of cold or iced water and making suction on or "stripping" the other extremity a continuous current is started.

Lenigallol (Pyrogallol Triacetate).—A white powder, insoluble in water, but gradually dissolved on warming with aqueous solutions of alkalies. Used in acute, subacute, and chronic psoriasis, and eczema in the same manner as pyrogallol. It does not affect the healthy skin and is practically non-poisonous.

Leukonychia.—A peculiar whitish discoloration of the nails due to the presence of air beneath them and in their substance. There are but four or five cases on record.

Leyden's Asthma Crystals.—These consist of sharply pointed, refractive, octahedral crystals, observed microscopically in the sputum of bronchial asthma.

Lignosulphite.—A liquid containing volatile oils, used for inhalation in tuberculosis.

Lipanin.—A purified olive oil, said to be prepared from the finest quality of virgin olive oil, in accordance with the formula of Prof. J. von Mering, and used as a superior substitute for cod-liver oil. It is more agreeable to the taste and smell than cod-liver oil; it emulsifies rapidly and is quickly absorbed, and it is acceptable in warm as well as in cold weather. It is said to keep better than cod-liver oil. Dose: for a child of one to six years, one to two teaspoonfuls twice daily; for older children, one-half to one tablespoonful daily; for adults, a tablespoonful two to four times a day. It has proven of great value in rickets (especially in combination with phosphorus), scrofula, phthisis, chronic rheumatism, anæmia, and during convalescence from protracted diseases.

Lipogenic Glycosuria.—The glycosuria of obese subjects, which, as a rule, is of minor significance and is rarely followed by a true diabetes.

Liquid Air.—As long ago as 1877, Pictet, by submitting oxygen to enormous pressure combined with intense cold, produced a clear bluish liquid which bubbled violently for a few seconds and disappeared as a cold white mist. This experiment proved oxygen to be not, as had been supposed, a permanent gas, but merely the vapor of a mineral. The first ounce of liquid air produced by Professor Dewar, some fifteen years ago, cost over three thousand dollars. Quite recently Tripler, of New York, has succeeded in producing it at the rate of fifty gallons a day at a cost of twenty cents per gallon. Dewar compressed nitrous-oxide gas and ethylene gas, and, by expanding them suddenly, produced a degree of cold which liquefied air almost instantly. Tripler applies the principle of utilizing compressed air, which is also a gas, for the production of the necessary cold. It is the coldest substance known, with the exception possibly of liquid hydrogen. A drop upon the skin has the same effect as iron at a white heat. Since, however, it does not burn, it seems admirably adapted for surgical uses, and is said to destroy pathological tissue much more safely than either caustic potash or nitric acid. The therapeutic uses of this powerful agent are still in their earliest infancy, but would seem to be very promising.

Liquid Hydrogen.—Produced in the same manner as liquid air under great pressure and the effect of cold. Its boiling point is 240° C. Its chief use at present seems to be in the manufacture of high vacuum tubes.

Liquor Carbonis Detergens.—This is a tar preparation which has been used with good results in chronic eczema. "It is prepared by taking 9 ounces (288.0) of tincture of soap-bark (quillaja-bark), and 4 ounces (128.0) of coal-tar, mixing and allowing them to digest for eight days, after which the mixture is filtered and used."—HARE. It should be used diluted in the proportion of 2 drachms to 4 ounces of water.

Listerine.—This is an agreeable, mild antiseptic containing, according to a published formula:

```
Oil of eucalyptus,
Thymol,
Oil of wintergreen,
Menthol.....................................āā gr. x.
Boric acid ....................................... ℥ ss.
Alcohol...........................................  ℥ ivss.
Water .......................................q.s. ad ℥ xvi.
```

On account of its deodorizing and antiseptic properties, it is largely used for the nose and throat in the form of washes and sprays. It may be given with water alone, 1 to 10, or in combination with boric acid, bicarbonate of sodium, iron, and various astringents.

Litten's Sign.—A constant visible movement of the diaphragm on inspection of the chest wall, described by Litten. It is an undulating movement or shadow beginning at about the sixth intercostal space, and descending on inspiration as a furrow at times as far as the costal margins. In expiration it returns to the starting-point. Seen best in the recumbent posture.

Little's Disease (Spastic paraplegia of infants, or diplegia spastica infantilis).—A disease due either to a birth palsy, the result of a severe and instrumental labor, or to a congenital defect of the cerebrum. Though present from birth, the disease is not noticed, as a rule, until the child begins to walk, that is, from the second to the sixth year. The first symptoms which attract attention are the rigidity of the limbs and the difficulty in walking. When the child attempts to walk he looks for aid and his awkwardness is increased. The limbs are rigid and the reflexes exaggerated. The attitudes when standing and walking are quite characteristic. There is talipes equinus of various degrees and a tendency to genu valgum. While standing and when walking the child drags one foot over and in front of the other. The gait is a waddling one, similar to that seen in congenital dislocation of the hip. The arms may or may not be involved. Occasionally they show a slight rigidity. There is no disturbance of sensation, and the functions of the bladder and rectum are normal. In some cases the intelligence is good; others show defects in memory and speech, idiocy, imbecility, strabismus, nystagmus, and epileptiform attacks.

Loeffler's Bacilli.—The diphtheria bacilli, at present usually designated as Klebs-Loeffler bacilli.

Loeffler's Solution.—See Toluol.

Loretin (a derivative of quinolin).—A yellow crystalline powder, without odor, used in surgery to prevent suppuration, also in eczema, lupus, and erysipelas. In solutions of two to five per cent it may take the place of carbolic solution; also used in five- to ten-per-cent ointment.

Losophan (Tri-iodo-cresol). — This drug consists of white needles soluble in ether and chloroform. It is used in eczema, sycosis, pityriasis versicolor, as an antiseptic and antimycotic. Best employed in one- to five-per-cent ointment or in alcoholic solution. Avoid this drug in acute inflammatory conditions.

Lucas-Championnière's Antiseptic Powder.—

R Powdered gray cinchona,
Powdered benzoin,
Iodoform,
Magnesium carbonate............Equal parts by volume.

Luschka.—A name associated with the pharyngeal tonsil at the vault of the pharynx, and known as Luschka's or the third tonsil. A small gland near the tip of the coccyx is also known by this name.

Lutaud's Lotion.—

R Chloral hydrate,
Tincture of eucalyptus.....................āā 10 parts.
Cocaine hydrochlorate........................ 1 part.
Distilled water.............................. 500 parts.

Used especially for pruritus vulvæ.

Lycetol (Dimethylpiperazin).—A white powder, soluble in water, and like piperazin, used as a solvent of uric acid. Also supposed to have diuretic properties. Dose, gr. x. to xv. three to four times a day, in gout and chronic rheumatism. This drug may now be obtained in the form of an alkaline carbonated mineral water, put up in bottles of 24 ounces.

Lysidin.—A new solvent for uric acid. It is a watery solution containing fifty per cent of a hygroscopic crystalline base, lysidum crystallizatum. Dose is ℳ xxx. to ℳ cl. in aerated water in divided doses during the day.

Lysol.—Almost a perfect antiseptic and disinfectant in solutions of from one-half to two per cent. A brownish, clear, oily fluid, with an odor of carbolic acid and creosote. With water it makes a clear, soapy solution, and acts as a lubricant. For the disinfection of instruments this latter fact is considered a disadvantage, for it renders them slippery; this may be obviated, however, by a subsequent washing in sterilized water. It is more powerful than carbolic acid and less poisonous. It has replaced creolin and carbolic acid, and bids fair to be used as extensively as bichloride of mercury, the indications for its use being practically the same. It is not altogether non-poisonous, and a few instances of poisoning—Cramer (*Centralbl. für Gynäkol.*, 1898, No. 39), Pourtales (*Archiv für Gynäkol.*, 1898, 57, No. 1)—especially after uterine irrigation, have been reported. In dermatology it is useful in dilution in pruritic affections, eczema marginatum, chronic eczemas, etc.

Mackenzie's Eye Lotion.—

℞ Corrosive sublimate................................ gr. i.
   Chloride of ammonium........................... gr. vi.
   Cochineal.. gr. iss.
   Alcohol .... ℨ i.
   Water... ℥ viij.
Mix, and filter for half a day.

Macrocheilia.—Hypertrophy of the lip. This may be congenital or secondary to acromegalia, myxœdema, strumous diathesis, chronic rhinitis, etc.

Magnesium-Nitric Test for Albumin in Urine.—

℞ Strong nitric acid .... ℥ iv.
    Saturated watery solution of sulphate of magnesia... ℥ xx.

Used in the same manner as Heller's nitric-acid test. It is said to be more sensitive and to give a sharper and more compact ring than the latter, and has the advantage of not staining the hands.

Malakin.—Yellowish, silky needles formed from salicylic aldehyde, and paraphenetidin. Used in acute articular rheumatism and in fevers. Dose, gr. xv. in cachets, four times daily.

Mallein.—Used as injection (1 mgm. to 1 cgm.) to detect glanders in the human subject as well as in veterinary practice.

Mal Perforant, Buccal.—A disease first described by Fournier, seen in instances of tabes and of late syphilis, and characterized by ulceration and perforation of the alveolar border of the upper jaw and the neighboring hard palate.

Malt Soup.—The following soup is used at the Breslau University children's clinic in gastro-intestinal diseases: Fifty grams of wheat flour are stirred into one-third of a quart of water (50° C.); to this 10 c.c. of a eleven-per-cent solution of potassium carbonate are added. The malt-extract mixture is then stirred into the mixture of milk and flour, and the entire mass is cooked together.

Malvina Lotion and Cream.—These are proprietary preparations which are said to have proven efficacious in freckles or lentigo. The lotion is said to consist of two grains of bichloride of mercury, three drachms of zinc oxide, and one pint of emulsion of almonds with rose-water. The cream, according to *New Idea*, consists of, approximately:

℞ Saxoline	℥ ss.
White wax	ℨ i.
Spermaceti	ℨ ss.
Bismuth	gr. xl.
Bichloride of mercury	gr. ss.
Spirits of rose (ℨ iv. oil to O i. water)	ℳ xx.
Oil of bitter almonds	ℳ i.

Heat the first three ingredients until melted, and while this is cooling add the bismuth and bichloride. When nearly cool add the perfume.

Mamarowsky's Method for Staining Skin Sections. —The sections are, according to the *Post-Graduate*, first cut in paraffin and then placed for twenty-four hours in saturated solution of bichloride of mercury, containing five per cent bichromate of potash and 0.6 per cent sodium chloride. They are then stained for fifteen minutes in a slightly heated picrocarmin, washed in water, and then stained again for thirty minutes in alum hæmatoxylin. The sections are next stained for one and one-half minutes in a saturated solution of picronitric acid, until the dark-red epidermis can be distinguished from the light rose-colored corium. Wash in water, dehydrate, clear, and mount. The horny layer and blood are stained yellow, the smooth muscle structure gold-yellow, the small-celled infiltration dark violet, the rete Malpighii violet. The staining is very permanent.

Man's Method of Treating Tænia.—A laxative is administered the day previous. The evening before wine and water may be drunk, but no milk. The following morning at eight o'clock, the patient is given in repeated doses the ethereal extract of filix mas (20 to 25 gm.) in capsule or pill, and as much wine and water as is desired. By eleven o'clock, as a rule, the worm is discharged.

Marie, Sign of.—A peculiar nervous tremor observed in exophthalmic goitre. The tremor may occur in all parts of the body, but is most frequently seen in the fingers and hands.

Markasol (Bismuth Boro-phenate).—An antiseptic used in surgical dressings.

Mastopexy.—Hypertrophy of the breasts.

Materna.—The name given to a new apparatus for modifying milk. It consists of a sixteen-ounce vessel with various graduations for milk, cream, water, and sugar of milk. By following the markings on the glass, the proper proportion of proteids, fat, and milk sugar for the various periods of childhood is said to be obtained.

Matzoon.—Milk which has undergone lactic-acid fermentation. Very palatable and used with good results in gastritis, fevers, nervous prostration, typhoid, and hyperemesis of pregnancy. That of Dr. Dadirrian is the best.

McClintock's Rule.—A pulse of one hundred or more beats per minute after childbirth indicates an impending post-partum hemorrhage. The physician should under no circumstance leave the patient's side until the normal pulse rate has been attained.

Medicated Soaps.—The method of applying curative agents by means of soaps has recently come into favor, especially since reliable firms have placed on the market really efficacious soaps of definite strength of medicament and in appropriate combinations. Almost all the newer drugs of dermatological worth can now be had in this form.

Medullin.—Spinal-cord extract, similar in preparation, uses, and dose to cerebrin.

Megalogastria.—An abnormally large stomach, without any functional or organic disturbance.

Megalonychosis.—A term used by Dr. E. L. Keyes, Jr., to designate a "universal non-inflammatory enlargement of the nails" (*Medical Record*, April 23, 1898).

Megalosporon.—One of the two distinct fungus plants causing ringworm. It belongs to the Botrytis. It gets its name from its forming, while fructifying, grape-like clusters. There are two subdivisions: (*a*) Megalosporon endothrix, and (*b*) Megalosporon ectothrix. According to Sabouraud the endothrix is never found in domestic animals; the ectothrix is present in dogs, cats, horses, pigs, and birds.

Melachol.—A strong solution of sodium citrophosphate, used in diseases of the liver.

Mel-Maroba.—A tonic said to contain the medicinal properties of manaca, caroba, stillingia, and iodide of potassium.

Melulose.—A name given by Hoff to a pure concentrated extract of malt.

Mercury, Asparaginate of.—Used in hypodermatic injections, passing rapidly into the circulation and being rapidly eliminated. One-half-per-cent solutions are said not to be painful. One centigram in a cubic centimetre of water may be repeated daily.

Mercury Colloidale.—See Hydrargyum Colloidale.

Mercury Sozoidol.—Antisyphilitic, especially as an injection:

    ℞ Hydrargyri sozoidol.......................... gr. xv.
    Potassii iodidi................................ gr. xxv.
    Aquæ destillatæ............................. ℥ iiss.
    M. S. Inject ♏ x. every day or every second, third, or fourth day.

Mercury Sozoiodolate.—Insoluble in water, soluble in salt water, used in powder form, one-per-cent ointment, and eight-per-cent lotion, for syphilitic affections.

Mercury Succinimide.—Used in hypodermatic injections as a form of mercury which does not precipitate albumin. Daily doses from gr. $\frac{1}{64}$ to $\frac{1}{30}$.

    ℞ Hydrarg. succinimid.......................... 1.30
    Aquæ destillatæ............................... 1,000.0
    M. S. One Pravaz syringeful.

Mercury Thymol Acetate.—As a hypodermatic injection, 1 part in 10 of oil. ♏ xv. injected once a week.

Merycism.—Another term for rumination in man. This is a very rare affection, and Sinkler (*Journ. Amer. Med. Assoc.*, April 9, 1898) has collected only thirteen cases in American literature. It is not a simple vomiting or regurgitation of food.

Methyl Salicylate.—Recommended by Lannois in the form of ointment with lanolin or vaseline for subacute and chronic articular rheumatism. It is advisable to cover the ointment with a bandage.

Methylene Blue.—First used by Mosetig for the treatment of cancer. Now used to test the eliminative power and permeability of the kidneys by the length of time it takes to appear in the urine. Also useful in gonorrhœa, cystitis, pyelitis, Bright's disease, and other affections. Has been used in gastric hyperacidity, migraine, neuralgia, and in malaria, when quinine is undesirable or has failed. Estay (*Médecine Moderne*, January 22, 1898) recommends it in diabetes mellitus. Best given in capsules or pills, with an equal quantity of powdered nutmeg. Dose is gr. i. to v. In dermatology it has recently been recommended by one of us in three-per-cent watery solution for intertrigo, eczema, etc.

Meunier's Sign of Measles.—A pronounced loss of weight day by day, noticed four or five days after contagion. This may reach 50 gm. daily, beginning five or six days before the occurrence of catarrhal and febrile symptoms.

Micajah's Wafers.—These are said to contain:

Mercury bichloride	gr. $\frac{1}{16}$
Zinc sulphate	gr. v.
Bismuth subnitrate	gr. xv.
Acacia	gr. v.
Carbolic acid	gr. iij.
Water	q.s.

Recommended by Dr. Juettner, of Cincinnati, in the treatment of rectal ulcers:

℞ Micajah's wafers 2
  Ol. theobrom. .... q.s.
  M. ft. supposit. iv.

Micro-Cautery.—A term employed to denote cauterization by the application of a fine-pointed instrument.

Microcidin (Sodium Beta-Naphtholate).—An antiseptic.

Microsporon Audouini.—A plant formerly supposed to be a trichophyton. It occurs in most cases of ringworm of the scalp in children, and consists of innumerable small round spores (3 μ in diameter). These spores usually lie in a dense mass.

Migrainator.—Dr. Sarason has suggested an instrument for the relief of migraine, which he describes in the *Deutsche med. Woch.*, No. 35, 1898. It consists of two plates, which are firmly pressed on the temples by a spring, and the device looks very much like a double truss. By the compression of the temporal arteries the circulation of the blood in the head is regulated and pain is relieved, especially in angiopathic hemicrania.

Migranin.—This is said to be a combination of antipyrin, citric acid, and caffeine in certain proportions. As its name implies, it is used for the paroxysms of migraine. Dose, gr. xv. every two to three hours, if necessary.

Migrol.—A salt composed of caffeine and guaiacetin. Used for nervous headache, neuralgia, and migraine, in doses gr. v. to vii. Said to be absolutely harmless.

Mikulicz's Disease.—A chronic enlargement of the lacrymal and salivary glands. Thus far it has not met with universal acceptance as a separate disease.

8

Millard's Fluid.—Used for the detection of albumin in the urine:

R Glacial carbolic acid (95 per cent)............... ℥ ij.
Pure acetic acid.......................... ...... ℥ vij.
Liquor potassæ................................. ℥ xviij

Mirror Speech.—A name given by Marcotte to a symptom observed in a girl twelve years of age, suffering from cerebral abscess, and consisting of an inversion of speech similar to that of mirror writing, so that sentences were uttered backward, *e.g.*, "Quille-tran-ser-lais-me-vous-lez-vous-te-tan-ma," when she wished to say: "Ma tante, voulez-vous me laisser tranquille."

Mirror Writing.—An inversion in writing so that letters and sentences must be reflected in a mirror to be read.

Mixed Toxins.—See Coley's Mixture.

Moebius' Symptom of Graves' Disease.—If the patient is told to keep his eyes fixed on your approaching finger, to converge them in other words, one or both eyes will wander toward the outer canthus. This is due to an insufficiency of convergence. This sign is by no means constant.

Mogigraphia.—A technical term for writer's cramp.

Mollin.—A superfatted soap said to contain about seventeen-per-cent excess of fatty matter. Used as a base for ointments, especially those of mercury and iodide of potassium.

Morrhuol.—An alcoholic extract of oleum morrhuæ, one part of which is said to represent thirty of the oil. Dose, ℳ iii. to xv. in capsules.

Morrhuol Creosote.—Small spherical capsules introduced by Chapoteaux, each containing three minims of morrhuol, the active principle of oleum morrhuæ, and one minim of creosote. Dose, four to ten capsules daily.

Morvan's Disease.—A necrotic disease of the fingers, existing only in Brittany and named after the physician who first described it. It is a chronic disease with cutaneous anæsthesia and painless whitlows. By some it is regarded as a form of syringomyelia; others consider it an infectious neuritis.

Motet's Operation for ptosis consists in cutting a strip in the superior rectus and pulling it through a buttonhole in the tarsal cartilage in order to suture it to the upper lid.

Murphy Button.—An appliance devised by Dr. Murphy, of Chicago, for visceral anastomosis without suture. It consists of two similar parts of mushroom shape with a canal running through the centre. The stem of one fits tightly in that of its fellow, bringing the bases into contact. Especially useful when necessity for haste is an important element in the case. When union has taken place the button is passed with the fæces.

Myasthenia Pseudoparalytica Gravis (Jolly).—An affection first described by Professor Jolly, of Berlin, and consisting of an abnormal muscular weakness in the legs and arms after using them for a short time. The eyelids and cheeks may also weaken after repeated winking and whistling. Chewing, talking, and deglutition are intact, thus distinguishing it from bulbar paralysis. The muscles show the "myasthenic reaction" of Jolly. This consists of a tetanic contraction of the muscle upon the application of the faradic current, which, however, becomes weaker upon the repetition of the irritation and finally disappears altogether.

Mydrin.—A mydriatic containing ephedine and homatropine. Used in ten-per-cent solution.

Myelopathic Albumosuria.—A term applied by Drs. Bradshaw and Warrington, of Liverpool, to a condition in which the persistent occurrence of albumose in the urine is associated with softening of the bones due to multiple myelomata.

Myophone.—An instrument devised by a French scientist to show that the nerves may remain alive many hours after death.

Myotonia Congenita, or Thomsen's Disease.—An hereditary affection, consisting of intention spasms, that is to say, tetanic contractions of the muscles upon voluntary movement. The muscles of the arms and legs are usually involved, rarely those of the face, eyes, and larynx. The muscles show the myotonic reaction of Erb, "the chief feature of which is that normally the contractions caused by either current attain their maximum slowly and relax slowly, and vermicular, wave-like contractions pass from the cathode to the anode."—OSLER. It is a family disease and lasts through life.

Myotonic Reaction of Erb is made up of normal mechanical, faradic, and galvanic excitability of the motor nerves, and an increased mechanical faradic and galvanic excitability of the muscles. Here with the galvanic current only closure contractions are obtainable, and these are as strong with the anode as with the cathode; the contractions are always slow, tonic, and prolonged.—Dr. GEORGE W. JACOBY.

Myrtill Pills (Jasper).—A pill devised by Dr. Weil, of Berlin, as a remedy for diabetes and used with some reported success. Each pill contains 0.12 cgm. ex. fol. myrtilli. From one to four pills three times daily is the ordinary dose.

Myrtol.—The essential oil of myrtle.

Naftalan.—Not to be confounded with naphthalin. A blackish-green mass, soluble in oils, ether, chloroform; insoluble in water. Parasiticide, analgesic, antiphlogistic. Useful in skin affections to replace tarry substances; in burns, erysipelas, rheumatism, and epididymitis.

Nail Sign in Malaria.—A slate-like discoloration of the nails which may be of value in the diagnosis of obscure febrile affections.

Nasal Antiseptics.—Dobell's solutions, Seiler's antiseptic tablets, Thiersch's solution, boric acid, glycothymolin, borolyptol, borofluorin, listerine, ichthyol, plasma nasal tablets, Douglass' tablets, boroformalin, alphasol.

Nasal Vertigo.—A dizziness similar to gastric and auricular vertigo, coming on after habitual nasal obstruction, traumatism, foreign body in the nasal fossæ, or subacute coryza.

Negro Lethargy.—This is a sleeping-sickness endemic on the western coast of Africa. A bacillus has been found and cultivated by Cagigal and Lepierre, of Coimbra. See "Sleeping-Sickness."

Neurexaresis.—A term applied to the extraction of a nerve by torsion.

Neurosin.—A term used for the glycerophosphates.

Nirvanin.—A new anæsthetic discovered by Professor Einhorn and used for infiltration anæsthesia; said to be non-toxic and a convenient substitute for cocaine. It may be sterilized without injuring the anæsthetic properties. Extensively investigated by Luzenberger (*Münchener med. Woch.*, January 3 and 10, 1899). It is said to be ten times less toxic than cocaine while possessing greater anæsthetic qualities. Employed in five-per-cent solution for hypodermatic injection.

Nitroglycerin.—Advocated by Flick in the treatment of hæmoptysis. It has also been combined with cocaine to counteract the tendency to collapse.

Nitrogen Gas forced into the pleural cavity by hydraulic power is a suggestion of J. B. Murphy, of Chicago, for the treatment of consumption. A tank of gas compressed to one hundred pounds pressure to the square inch is connected with a graduated gas bottle, and the amount introduced is regulated.

Nosophen.—A product of iodine on solutions of phenolphthalein, containing sixty-one per cent of iodine. A grayish-yellow powder, odorless, tasteless, insoluble in water but more soluble in ether and chloroform. Recommended for insufflation in acute coryza, chronic rhinitis with hypersecretion, balanitis, moist chancre, eczema, etc.

Nucleohiston.—A proteid found by Jolles in the urine of patients suffering from pyelitis, cystitis, pyelonephritis, etc. This substance contains phosphorus which may be freed by treatment with hydrochloric acid.

Nutrose.—A proprietary nutrient preparation from the casein of milk, to which sodium is added. It is a color-less, tasteless powder, readily soluble in water. It is indicated in convalescence from protracted diseases.

Nymphfibulation.—Some years ago Dr. Collier, of New York (*Medicine*, vol. iv., p. 480), reported to the New York Academy of Medicine the case of an Austro-German woman in whose labia a husband had bored holes. Through these a padlock was locked during the absence of the husband. This custom was once quite the mode in France and Germany, and was the subject of a satiric poem by Voltaire. It evidently survives in Russia. Hinrichsen (*Vratch*, April 15, 1898) reports the case of a Russian woman in whose labia majora her husband had pierced holes and closed these by a gold locket, the key of which he kept always in his possession. In the Hôtel Cluny Museum in Paris is to be seen a similar lock said to have been employed by an historical character while away at the wars.

Oblique Heredity.—The same as Indirect Atavism.— SEDGWICK.

Oenilism.—A form of alcohol intoxication with digestive and nervous disturbances.

Oliguria.—Diminished excretion of urine. Observed in fever, emphysema, myocarditis, endocarditis, and pericarditis, acute nephritis, hypertrophied prostate, atony of the bladder, etc.

Oniomania.—A mania for making purchases.

Onychophagia.—The habit of nail-biting.

Oophorectomy.—Recently advocated for treatment of inoperable carcinomata of the breast and uterus. Cheyne reports two cases with satisfactory results.

Oophorin.—Used by Landau, of Berlin, in the treatment of climacteric disturbances in doses of gr. iv. to viii. Amenorrhœa, dysmenorrhœa, and chlorosis of ovarian origin have responded to its use. Jayle (*Rev. de Gynœcol.*, August, 1898) reports favorable results with ovarian extract. Senator, Rossier, Fosbery have also reported favorably.

Opalisin.—A fourth proteid found in milk by Wroblewski. It is said to exist in large quantities in human milk. It is obtained by the addition of sodium chloride to the fluid remaining after the precipitation of the casein in human milk by hydrochloric acid.

Ophthalmoplegia.—A disease due to atrophy of the nuclei of the nerves which move the eyeball. May be (*a*) externa or (*b*) interna. In the former there exists ptosis, strabismus, nystagmus, or diplopia; in the latter there is loss of pupil reflex to light (locomotor ataxia and general paresis).

Opotherapy.—Same as Organotherapy.

Orexin.—A white powder, soluble in hot water, of a bitter and burning after-taste. In doses of from gr. ii. to

iv. it is useful in anorexia, improving the appetite and general condition. It also acts as an anti-emetic in the vomiting of pregnancy. The existence of any gastric disease proper is a contraindication to its use. Attacks of syncope and tinnitus aurium have been observed after the administration of this drug.

Orexin Tannate.—Stomachic, appetizer, and antiemetic. A yellowish-white powder, odorless and practically tasteless, insoluble in water, but easily soluble in acids, especially hydrochloric. The dose for a child is gr. v. twice a day.

Organotherapy.—The same as the treatment by animal extracts.

Orthoform.—A fine white powder, barely soluble in water, and not hygroscopic. Mostly used in surgery on account of its anæsthetic and antiseptic properties. Its analgesic properties are shown by the fact that the nerve filaments with which it comes in contact are rendered insensible. It is of special value in wounds, burns, ulcers (syphilitic, tuberculous, carcinomatous, and particularly those following vaccination). By its use the discharge from wounds is diminished. It may be used in the form of powder, gauze, or ointment. In solution it has been used as a bladder wash for tuberculous cystitis. For internal use it has been tried in ulcer of the stomach in doses of 0.25 to 0.5. The results thus far have been very satisfactory, the diminution of pain being its most marked effect. Applied after the extraction of teeth it acts as a good local anæsthetic. The greatest objection to the use of this drug is its expense.

Ossalin.—A neutral grayish fat with tallow odor, made from fresh bone marrow. Like lanolin it is hygroscopic. An ointment base.

Osteo-Arthropathie Pneumonique.—The term employed by Marie to denote hypertrophy of the end phalanges as seen in cases of chronic pulmonary and heart lesions.

Otomyasthenia.—A weakness of hearing due to asthenia of the ear muscles, tensor tympani and stapedius.

Otomycosis.—The growth and development of the aspergillus in the auditory meatus.

Ouabain.—White crystals, soluble in hot water, used in asthma and whooping-cough for its action on the respiratory centres. Dose, gr. $\frac{1}{1000}$.

Ovarin.—Extract of ovaries of the cow. Dose, gr. iii.

Oxyphenylsulphonic-Acid Test for Albumin.—

R Acidi oxyphenylsulphonici...................... 3 parts.
Acidi salicylsulphonici...................... ... 1 part.
Aquæ .... 20 parts.

One drop of this solution added to 1 c.c. of urine gives, according to Bourceau (*La Médecine Moderne*), a white transparent precipitate of albumin. Peptone, propeptone, alkaloids, urates, and phosphates are not precipitated.

Pagenstecher's Ointment.—The yellow oxide-of-mercury ointment, which is used so extensively in ophthalmology, especially for blepharitis ciliaris.

R Hydrargyri oxidi flavi .... gr. i.–gr. iv.
Vaselini .... ℥ ss.

Pacini's Solution.—Used with the Thoma-Zeiss apparatus for the counting of red blood cells.

Bichloride	1
Sodium chloride	2
Glycerin	100
Distilled water	300

Or—

Bichloride	1
Sodium chloride	2
Distilled water	200

Panas' Operation for Strabismus.—After opening the conjunctiva by a horizontal incision, the tissues are thoroughly divided about the sheath of the muscle, in the usual manner. The muscle is then taken up on the hook and pulled well outward, so that the inner margin of the cornea is on a line with the external canthus. The tendon is then thoroughly divided, the wound closed by a suture, the operation repeated on the other eye, and both eyes are bandaged for twenty-four or forty-eight hours.

Pancreatin.—A mixture of the ferments naturally contained in the pancreas of warm-blooded animals. It has been used as a substitute for pepsin in dyspepsias of all kinds. It is given in powder form as such, or in combination with pepsin, bicarbonate of sodium, bismuth, orexin, etc. The dose is gr. v. to x.

Panopepton.—A brown liquid sold as a predigested food. Prepared from lean beef and whole wheat cooked, peptonized, dried, and preserved in sherry. Dose, one teaspoonful to a tablespoonful.

Panophthalmitis.—A general inflammation of the entire eyeball.

Papain or Papayotin.—A peculiar ferment obtained from the juice of papaw, the fruit of an herbaceous tree (Carica papaya) cultivated in tropical countries. As it is sold, it appears as a grayish, fine powder, with an odor and taste suggestive of pepsin. It has the power of dissolving fibrin and muscular fibres, and its advantage over pepsin is said to lie in the fact that it acts in acid, alkaline, and neutral media. The latter fact, however, is disputed by Brunton and Martin. It has been used in dyspepsia and gastric catarrh, in doses of gr. v. to x. Externally it has been used in chronic eczema and warts.

Papine.—A brown liquid sold on prescription as an anodyne, and said to contain a purified opium.

Papainproteolysis.—Papain, which is a proteolytic enzyme of the papaw plant, has been found by Chittenden capable of converting large quantities of the various proteids into true peptones.

Parablast.—A dermatological dressing brought forward by Unna. A thick-webbed tissue spread with lanolin, damar, caoutchouc, and resin, to which any remedial agent can be added.

Paraacetphenetidin.—Phenacetin or phenetidin.

Paraform.—Obtained by heating an aqueous solution of formaldehyde. It consists of a white, crystalline body, insoluble in water and used as an intestinal antiseptic. Dose, gr. ii. to viii. daily.

Paraform Pastilles.—Tablets used for generating formic aldehyde gas.

Paraldehyde.—A hypnotic produced by the action of

sulphurous gas or hydrochloric acid on aldehyde. It is a colorless liquid, with an ethereal smell and a very disagreeable taste, soluble in water. It is said to be superior to opium and chloral as an hypnotic, but inferior as an analgesic. Its disadvantages are the unpleasant taste and odor, and its tendency to upset the stomach. Dose, ℥ xx. to 3 i., preferably in capsules. May be given per rectum. It is relatively non-poisonous, large doses causing death by respiratory paralysis. Peabody recommends it in thirty-drop doses for bronchial asthma. Relief is said to be prompt in three doses, relieving the dyspnœa and permitting the patients to sleep.

Parataloid.—Another name for tuberculin.

Parotid-Gland Therapy.—R. Bell (*Klinische therapeut. Wochenschrift*, May 29, 1898) has given the dried parotid gland of sheep for ovarian disease, and has reported sixty instances of enlarged and painful ovaries in which this medication effected a cure. The dose is gr. v. three or four times a day.

Pavor Nocturnus.—Synonymous with the night terrors of children, and due to adenoids, enlarged tonsils, rhinitis, dyspepsia, and anæmia.

Pavy's Disease.—A name applied to periodic, paroxysmal, or cyclic albuminuria. This condition occurs for the most part in nervous and irritable young persons, and consists in the appearance of albumin in the urine at some time of the day and its absence at other times. It may appear in the morning and be absent in the afternoon. Worry, mental strain, and physical exertion appear to cause its appearance in the urine. In a number of cases it is a functional disturbance, and tends to get well; in others it paves the way for grave organic kidney changes. This intermittent appearance of albumin is occasionally observed in chronic nephritis.

Pellotin.—A hypnotic recently introduced by Jolly, of Berlin. Used in locomotor ataxia. Dose, gr. $\frac{1}{3}$. Gr. $\frac{1}{8}$ subcutaneously has produced dangerous collapse.

Pentose (Salkowski).—This term is applied to a group of carbohydrate substances—arabmose, rhamnose, xylose —which occur in normal and pathological urine, and which are of interest because they react with Trommer's, bismuth, and phenylhydrazin tests, and therefore may be mistaken for glucose. These substances, however, do not ferment with the yeast test and do not polarize to the right.

Peptenzyme.—Said to be prepared from the various digestive animal glands and the ferment extract of the spleen and liver, slightly benzoated and mixed with sugar of milk. It is claimed to contain the enzymes of these organs in the same state in which they exist in nature, and to act as a digester of all forms of food. It may be given in powder or tablet form before or after meals.

Peptothyroid.—A peptonized preparation of thyroid extract used by Maurange for conditions in which thyroid is indicated. It is claimed that the by-derangement is less with this medication.

Peptovarin.—Peptonized ovarian extract. Advocated and used by Maurange.

Perco.—Another term for perucognac.

Peritoneopexy.—Fixation of the uterus by the vaginal route, in the treatment of retroflexions of this organ, by a procedure recommended by Gottschalk. A transverse incision is made in the anterior vaginal wall at the insertion of the neck of the uterus, just as in vaginal hysterectomy; the peritoneal cavity is opened by a free incision of the vesico-uterine cul-de-sac in a transverse direction. For the steps of the operation, see *Centralblatt für Gynäk.*, No. 4, 1899; or *La Presse Médicale*, February 15, 1899.

Peritonism.—A neologism introduced by Gubler, originally applied to forms of pseudo-peritonitis in which the symptoms were present without the essential elements of the disease. It is observed chiefly in neuropaths, and in most cases this simulation of peritonitis is associated with hysteria, while some slight lesion of the appendix, a tube, or of an ovary, or a floating kidney may be present.

Perlèche.—This is an infectious and contagious disease affecting infants and children—for the most part in institutions—and consisting in the formation of painful fissures at the commissures of the lips. It must be differentiated from syphilis by the absence of all other symptoms. The treatment consists of careful attention to all the material connected with nursing (bottles, nipples, etc.), and the application of nitrate of silver, alum, permanganate of potassium, or five-per-cent solution of methylene blue.

Peronin (hydrochlorate of benzoyl-morphine).—A fine, whitish, odorless powder, with a bitter taste, very soluble in water. Used by Eberson for bronchitis, pulmonary tuberculosis, and whooping-cough. A narcotic which takes its place between codeine and morphine, producing more profound and calmer sleep than the latter, without being preceded by phenomena of excitation. Dose, gr. $\frac{3}{20}$ to $\frac{5}{10}$, t.i.d., for adults; and to children, gr. $\frac{15}{1000}$ for each year of their age.

Pertussin.—As its name implies, this is a remedy for whooping-cough. It is prepared by mixing the fluid extract of thyme with syrup, so as to procure an infusion in the strength of 1:7. Prof. Ernst Fischer (*Deutsche med. Wochen.*, November 28, 1898) used this drug after having failed with tussol. He reported that the drug was pleasant to take, and that in a short space of time the disease was changed into almost an ordinary bronchitis. The attacks seemed to become less frequent and milder, the phlegm looser, and the cyanosis practically disappeared. This drug also proved of service in acute and chronic bronchitis, emphysema, and in the bronchitis following anæsthesia.

Pertussis, Bacillus of.—Koplik, Czaplewski, and Hensel, each working independently, isolated and described almost simultaneously a bacterium not found in other sputum. It is characterized by a minuteness similar to that of the bacillus of influenza.

"If stained with Loeffler blue it appears as an exceedingly minute, delicate, thin, short bacillus form, much thinner than the diphtheria bacillus, and not more than one-third to one-half its length. It measures 0.8 to 1.7 μ in length, and 0.3 to 0.4 μ in breadth. In pure culture it is not decolorized by Gram stain."

Perucognac (Perukognak).—A preparation containing balsam of Peru, advocated by Schmey and others in the treatment of tuberculosis. It contains the active principle of 25 gm. of balsam of Peru with ten per cent. cinnamic acid, in 1 litre of cognac.

Pfeiffer's Bacillus.—Bacillus of influenza.

Pfeiffer's Disease (Glandular Fever or Drüsen-Fieber).—Characterized by a sudden onset and an enlarged, smooth swelling on one side of the neck, due to enlarged lymph nodes. Twenty-four to forty-eight hours afterward the opposite side becomes similarly involved. The swellings increase for from four to seven days, remain stationary for a brief period, and subside in from two to three weeks.

Phacotherapy.—Another term for heliotherapy.

Phagotherapy.—A treatment by superalimentation, especially in phthisis.

Phakoscopy.—A procedure of personal inspection of the media of one's eye. It is accomplished by looking through strong myopic glasses (40 diopters) at a candle flame placed at the back of a dark cabinet. The parallel rays throw upon the retina a shadow carried from the lens, which is thus seen by transparence. The slightest opacity is thus made apparent to the retina in the form of a dark spot, a streak, or a star.

Phenalgin.—Said to be an ammonio-phenylacetamid. Of occasional use as an antipyretic and analgesic in dose of gr. iii. to x.

R Phenalgin.. gr. iij.
  Quininæ sulph .... gr. ij.
M. S. One powder every three hours.
                                    —HOFHEIMER.

9

Phenatol.—A compound of acetanilid, sodium bicarbonate, carbonate, sulphate, chloride, and caffeine.

Phenazonum.—Phenazone is the official name in the British Pharmacopœia for antipyrin. The drug is still unofficial in the United States Pharmacopœia. It is more or less decomposed or thrown down in solutions with a variety of chemicals, hence should be given alone or be cautiously combined.

Phenedin.—Same as Phenacetin.

Phenocoll Hydrochlorate.—Antipyretic, nervine, and antirheumatic. Dose is gr. viii. to xv., in powder form. Has been recommended for whooping-cough.

Phenol-Phthalein.—A solution used for testing the total acidity in a given specimen of gastric juice.

Phenopyrin.—An oily liquid containing equal parts of phenol and antipyrin.

Phenosol.—A product of salicylacetic acid and phenetidin, containing fifty-seven per cent of phenacetin and forty-three per cent of salicylic acid. Used in acute articular rheumatism in doses of gr. viiss., two to six times daily.

Phenylacetamid.—See Acetanilid.

Phonendoscopy.—A method of stethoscopic examination by the means of an instrument devised by Bianchi, called the phonendoscope. The heart sounds may be heard through the clothing, but to appreciate distinctly the sounds of circulating blood in the vessels and muscular fremitus, the disc is applied directly to the skin. The various organs can be mapped out by the shades of tone communicated to the ear.

Phototherapy.—Light as applied to the cure of variola by the application of hangings to exclude all but the red rays (Finsen); of rubeola (Chatinière). All light is here excluded from the room, excepting that supplied by a photographic lantern. Good results are reported (*Presse Méd.*, September 10, 1898).

Phthinoid Bronchitis.—A form of chronic bronchitis which bears a close resemblance, as far as its symptoms are concerned, to pulmonary consumption.

Phytolin.—A liquid containing the active principle of phytolacca decandra. Used in obesity, etc. Dose, ℥ x.

Pichi.—A shrub found in Chili, South America, containing an active principle which has proven of worth in chronic cystitis, enlarged prostate, diseases of the liver, gall stones, and uric-acid diathesis.

Picric Acid.—A yellow fluid, lately used with excellent results, in the form of a one-per-cent solution, for the treatment of burns of the first and second degree, acute eczema, dermatitis venenata, and chilblains. For this purpose Esbach's reagent may be used. Its most marked effect is the almost immediate relief of pain. Its disadvantage is the staining of everything with which it comes in contact. Stains may be removed from the clothing by boiling, and from the hands by a subsequent washing in alcohol or in a saturated solution of carbonate of lithia. The authors have found this fluid very efficacious in relieving the burning sensation following the local application of carbolic and trichloracetic acids, nitrate of silver, and formalin.

Picrol.—A fifty-two-per-cent iodine antiseptic.

Pilonidal Sinus.—This term was applied by Dr. R. M. Hodges, of Boston, to a suppurating sinus, not usually mentioned in text-books, containing hair, and situated in the coccygeal region. An abscess may form when the sinus becomes closed. To effect a cure, the small tuft or ball of hair at the bottom of the sinus must be extracted. Since it is not congenital and said never to occur until puberty, the term dermoid cyst is inappropriate.

Piperazin (also known as piperazidin). — This drug consists of colorless crystals, which have a faint odor, pungent taste, and which are hygroscopic. It is freely soluble in water. When brought in contact with uric acid, it forms a soluble uric-acid salt—the urate of piperazin; on this account it has been recommended in place of the lithium salts for uric-acid diathesis, renal and vesical concretions, and ever for renal calculi. It is said to be one of the best uric-acid solvents known. The dose is from gr. ii. to viii., three or four times daily. It has been given subcutaneously in five-grain doses. This drug is also sold in the form of a piperazin water, thus combining two great factors in the treatment of uric-acid conditions.

Piperine. — Crystalline prisms, having a formula isomeric with that of morphine. A febrifuge in doses of gr. ii. to v.

Pixol.—A syrupy liquid, prepared by adding to one part of green soap and three of tar a solution of caustic alkali. A cheap general disinfectant.

Plasma Nasal Tablets.—A proprietary tablet composed of

℞ Sodium chloride.................................. gr. 5¼
Sodium sulphate................................. gr. 1½
Sodium phosphate .... gr. ¼
Potassium chloride............................. gr. ⅖
Potassium sulphate .... gr. ¼
Potassium phosphate .... gr. ⅓
Each tablet to be placed in two ounces of water.

Plaster Muslins (Pflaster-Mull). — Dermatological dressings having gutta-percha and oleate of alum as a basis, and containing the desired remedy.

Pneumaturia.—A term applied to urine which bubbles and contains a considerable quantity of gas when passed. Frisch (*Wien. klin. Wochen.*, No. 39, 1898) has published a case of thrush of the bladder in which this symptom was prominent.

Pneumochemic Treatment. — Inhalations of medicated vapors, useful in pulmonary affections, especially bronchorrhœa and irritable coughs of phthisis.

Pneumococcia.—A generalized infection by the pneumococcus, usually secondary to pneumonia.

Poikilocytosis.—A condition in which the blood corpuscles are of various sizes and shapes. Seen in severe cases of chlorosis, and in pernicious anæmia, leukæmia, etc.

Polyform.—A preparation sold under the name of Edison's polyform composition. Said to contain (A. P. A.):

℞ Chloroform .... ℥ ij.
  Ether .... ℥ i.
  Alcohol .... ℥ iss.
  Chloral hydrate................................. ℥ ij.
  Camphor .... ℥ i.
  Morphine sulphate............................. gr. vi.
  Oil of peppermint .... ʒ i.

Polyphagia.—An excessive appetite (see Bulimia).

Polyuria.—Marked increase in the excretion of urine. Observed in diabetes mellitus and insipidus, chronic interstitial nephritis, amyloid degeneration of the kidney, neurasthenia, after the absorption of exudates (pleurisy, ascites, etc.), and after infectious diseases.

Polyvalent Serums. — Serums derived from animals infected by a number of different streptococci, which are hoped to find a field of usefulness in instances of accidental infection by some unknown streptococcus. It is questionable whether an efficacious serum of this kind will ever be produced, corresponding to the monovalent serums now in use.

Porencephalus.—An atrophy or absence of brain substance, with a resulting cavity which may extend into the ventricle.

Potamophobia.—A morbid fear of rivers and water in general.

Potassium Permanganate. — Recently advocated in morphine poisoning, if given early; practically useless when the alkaloid has been absorbed. Much thought of in solution for urethral irrigation in gonorrhœa, in strengths of from 1:5,000 to 1:1,000. Two- to five-percent solutions have been recommended by Dombrowsky

(*Semaine Médicale*, October 15, 1898) for fissured nipples. Recommended internally by Ringer for atonic amenorrhœa. Useful in poisoning by phosphorus, hydrocyanic acid, muscarine, colchicine, and rattlesnakes. In the latter instance the part may be bathed in a solution or the fluid be injected in a circle around the point of infection.

Prochoresis.—The motor function of the stomach.

Proctoscope.—Kelly's instrument for examining high up in the rectum is considered by many a superior instrument to the ordinary rectal speculum. A strong reflected light is required.

Proptosis.—The prominence of the eyeballs, as seen in Basedow's disease.

Prostatic Extract.—This has been used by Oraison (*Gazzetta degli ospedali e delle cliniche*, May 19, 1898) for prostatic enlargement. Of the seven patients treated he claims five cures, one amelioration, and one failure. The dose is 3 to 12 grams of the powdered gland in pill form, or ℨ ii. to ℥ i. of the glycerin extract.

Protargol.—One of the many substitutes for silver nitrate which has appeared within recent years. It is a silver proteid compound, and consists of a yellowish powder, easily soluble in water, and not precipitated by albuminoid substances, acids, or alkalies, nor reduced by the action of light. It has been used quite extensively in gonorrhœa of both the male and female, in strengths of from one to two per cent. It is said to be less irritating than the nitrate of silver, to have a greater power of penetration, and to "dry up" the discharge more quickly. It has been used extensively in eye practice for purulent affection, dacryocystitis, trachoma, etc., in strengths of from ten to twenty per cent. It certainly appears to be less irritating and smarting to the conjunctival mucous membrane than silver nitrate, and is not so liable to cause deposits on the cornea.

Protonuclein.—The primal nuclein of individual tissue cells. Obtained by mechanical separation from the glands and tissues containing it. For medicinal use it is preserved by drying and molecular investment with benzoin. It is claimed for it that it increases the production of white blood cells, and hence enhances the reparative forces of the economy. In chronic conditions, gr. vi. to xii. in powder or tablet, before each meal and at bedtime. It should be masticated and dissolved in the mouth before being swallowed. Indicated in neurasthenia, anæmia, malarias, and general wasting diseases. Used externally in powder form for ulcers, fistulæ, fissures, etc.

Pruritus (Bath).—The term bath-pruritus has been applied by Stelwagon (*Philadelphia Medical Journal*, October 22, 1898) to a form of burning or itching experienced by some persons immediately after a bath. The sensation varies from a slight pricking to an intense and almost intolerable itching, which is usually aggravated by attempts to scratch or rub the skin violently. It is usually situated in the lower extremities, sometimes in the upper, and is occasionally general. The attack lasts from a few minutes to half an hour or more, being of longer duration when the patient goes directly from the bath to his bed.

Pseudo-Cyesis.—Spurious pregnancy. Observed in hysterical women who have passed the menopause.

Pseudo - Erysipelas.—An erysipelas-like affection, starting from the site of vaccination, of self-limited extent and duration, not going beyond the elbow and shoulder-joints (Allen).

An erysipelatous condition of the face, of limited duration, without marked constitutional symptoms (Bulkley).

Pseudo-Tetanus Bacillus.—Tavel reports a bacillus resembling that of actinomycosis, and at times the tetanus bacillus. Like the tetanus bacillus, it carries its spores at the end, but they are more egg-shaped; it has twelve flagellæ, while in that of tetanus they are more numerous. It does not grow on gelatin. It is found in perityphlitis and appendicitis, and is thought to lend its odor to the pus of perityphlitic abscesses.

Psittacism.—Relating to the affection psittacosis.

Psittacosis.—In 1897 Dupuy described an epidemic disease which occurred in bird dealers, especially those having charge of parrots. The disease sets in with pro-dromal symptoms of loss of appetite, depression of spir-its, malaise, and muscular weakness; this is followed by diarrhœa, bad-smelling stools, fever, and convulsions. At the height of the disease the patient looks like a typhoid case. In severe instances pneumonia complicates the condition. The disease is infectious and contagious, being carried from animal to man and from one person to the other. The specific bacterium, which has been found in the blood by Nocard, resembles biologically the typhoid bacillus and the bacillus coli communis. The prognosis is bad, death taking place in from eight to twenty days.

Psychopathia Chirurgicalis.—A term used by Pryor to describe a mania for being operated upon, possessed by certain patients, especially women.

Ptomatinuria.—The presence of ptomaines in the urine, seen in a great variety of febrile and gastro-intestinal disturbances.

Pulmonin.—A fresh extract of calves' lungs, given usually with guaiacol carbonate in tablet form.

Pulsus Paradoxus (Paradoxical Pulse).—Entire absence or weakness of the radial pulse during inspiration. Observed most frequently in concretio pericardii cum corde, or adherent pericardium. Occasionally seen in mediastinitis, mediastinal tumor, and stenosis of the bronchi.

Pyloric Stenosis (congenital).—A number of these cases have recently been reported by Rolleston, Finkelstein, Schwyzer, Melzer, Ashby, and others. The condition is present from birth, and usually proves fatal within a few weeks. The affection is rare, only seventeen cases having been thus far reported.

Pyoktanin.—The commonly used German term for pyoctanin or methyl violet. According to Liebreich, a mixture of several methyl violets. Yellow pyoktanin is used in ophthalmic surgery.

Pyramidon.—A new form of antipyrin. Has been recommended as an antipyretic, analgesic, and antirheumatic, and found of service in tuberculosis, rheumatism, migraine, trigeminal neuralgia, sciatica, alcoholism, neuritis, lumbago, etc. In the headache of psychoses, Laudenheimer (*Therapeut. Monatshefte*, 1898, Heft 4) has found it very efficacious. Dose, 10 to 15 gm. three times a day.

Pyrantin.—A new antipyretic obtained by the action of anhydrous succinic acid on phenetidin. It is soluble in alcohol, and combines with sodium hydrate (soluble pyrantin). Clinical observations are as yet wanting.

Pyretometer.—A term recently introduced by the Randall-Faichney Company, of Boston, to take the place of clinical thermometer, fever thermometer, etc.

Pyridin.—A colorless, volatile liquid, having a penetrating odor, made from Dippel's oil, and employed for inhalation in asthma, emphysema, and to overcome sensations of oppression in the chest. A drachm may be allowed to evaporate in a saucer near the patient, or fifteen drops or more may be inhaled from a handkerchief three times a day. This has been used in gonorrhœa as an injection:

R Pyridin .... gtt. viij.
Aquæ .... ℥ iv.

Pyrosal.—Used in acute rheumatic affections. A synthetic compound of salicylacetic acid and antipyrin, containing about fifty per cent of antipyrin and thirty-seven per cent of salicylic acid. It is with difficulty soluble in water, ether, and alcohol, but decomposed readily by acids and bases into its constituents. The dose is gr. vi. to viii., two to six times a day.

Pyrozone.—A proprietary preparation, being a fifty-per-cent solution of peroxide of hydrogen in sulphuric ether. Put up in small glass tubes, which must be opened with care to prevent explosion. A valuable caustic in the removal of warts, pigmentations, etc., and has been used successfully by one of the authors in the treatment of lupus erythematosus.

Quanjer's Method for the expulsion of tænia.—At 7 A.M. 35 to 40 gm. of aq. laxat. viennensis are given. Then 5 gm. of filicin, divided into eight or ten parts, are administered in soft gelatin capsules. The first two capsules are given at eight o'clock, and then two more every ten minutes until all are taken. A sip of port wine or madeira may be taken to prevent nausea. If the bowels have not moved by ten or eleven o'clock, another dose of the laxative is given.

Quinaseptol (Merck).—An internal antiseptic, particularly adapted for the genito-urinary apparatus, not being decomposed by the urine.

Quincke's Spinal Puncture.—This consists of a puncture of the spinal cord by means of a needle inserted between the second and third or third and fourth lumbar vertebræ. The result shows the presence or absence of fluid, the character of the same (purulent or serous), its pressure, and the existence of tubercle bacilli can be determined. It is of value in differentiating tuberculous meningitis from the cerebro-spinal form, and by relieving the intracranial pressure it at times ameliorates the severe symptoms. Puncture is to be made, of course, only under the strictest antiseptic precautions.

Rabies.—A blastomycete has been discovered by Memmo, which he thinks pathogenic and specific, having detected it in the blood of four dogs dying of rabies and in animals inoculated.

Radiography.—Same as skiagraphy. A term used more commonly in France.

Rag-Sorter's Disease.—An affection observed among workers in paper factories, and presenting symptoms which closely resemble internal anthrax. It is presumably a bacillary disease, and the bacillus proteus hominis capsulatus has been found in the internal organs.

Railway Brain.—Traumatic hystero-neurasthenia; very similar to railway spine.

Reader's Cramp.—A condition to which E. W. Wright has called attention, and which consists of a spasm or cramp of the ocular muscles following prolonged reading.

Red Bone Marrow.—Extract of the fresh or desiccated marrow of the large bones of oxen and sheep. Glycerin preparations are also used, among which there is one called carnogen. It has been used successfully in purpura, leukæmia, pseudo-leukæmia, psychoses, and other conditions in which iron would usually be indicated.

Resol.—Similar to creolin.

℞ Tar	10 parts.
Caustic potash	2 "
Methyl alcohol	2 "

Resorbin.—An ointment base of almond oil and wax, emulsified by water with gelatin and soap added.

Resorcinol.—A combination of iodoform and resorcin.

Rest Treatment.—Bed treatment, which has been so efficacious in neurasthenia and various forms of nerve ailment, has been advocated by Serieux in acute melancholia, as especially efficacious. The symptoms which seem particularly to indicate its use are cerebral anæmia, cyanosis, œdema of the extremities, neurasthenic symptoms, emaciation, chlorosis, stupor, suicidal tendencies, automutilation, insomnia, circulatory and respiratory disorders, digestive disturbances, and low temperature.

Retinol.—This is an antiseptic, the product of the dry distillation of colophane. It is used as a vaginal application on tampons, and as a vehicle for various antiseptics, such as resorcin, ichthyol, aristol, etc. It has been used as an injection in gonorrhœa, and also has been given as a balsamic in from ℨss. to iss. per day.

Retrocalcaneal Bursitis.—Another term for achillodynia. An inflamed bursa is not the only etiological factor, since flat-foot, gonorrhœa, gout, rheumatism, syphilis, caries of the os calcis, and certain occupations may also cause achillodynia.

Reusner's Sign of Early Pregnancy.—As early as the fourth week of pregnancy, a marked increase in volume of the pulsation of the uterine arteries may be felt per vaginam in the posterior cul-de-sac.

Rhabdomyoma.—A rather rare form of muscular tumor (myoma), which contains striped muscular fibre.

Rhinitis Tablets.—A compressed tablet which is occasionally very efficacious in acute coryza. It consists of—

℞ Camphor... gr. ¼
Powd. belladonna................................... gr. ⅛
Quin. sulph... gr. ¼
Dose : One tablet every half-hour until four are taken, then one every hour until throat feels dry.

Rhinomiosis.—A term used by Dr. Jacques Joseph, of Berlin, to indicate operative shortening of the nose.

Rhizomelic Spondylosis.—Ankylosis of the spine and of the limbs where they join the trunk (P. Marie, *Rev. de Méd.*, April 10, 1898). The spinal ankylosis is associated with bony outgrowths from the vertebræ, and is most marked in the lumbar region. The spine is fixed in flexion, resulting in considerable kyphosis. The hip is more markedly affected than the shoulder-joint, and is the only one in which true ankylosis exists. There is considerable limitation of motion in the shoulder-joint. Walking depends on movements of the knees and ankles.

Riga's Disease.—A disease first described by Fede, consisting of an hyperplastic condition of the under surface of the tongue. The disease is rarely seen outside of Italy.

Rinman's Sign of Early Pregnancy.—In two cases, slender cords radiating from the nipple were observed; these were considered hypertrophic acini of the glands. This observation has been confirmed by others.

Robert's Modification of Heller's Albumin Test. —See Magnesium Nitric Test, page 107.

Romberg's Symptom.—A swaying to-and-fro movement when the patient stands with feet together and eyes closed. Seen very frequently in tabes dorsalis, and occasionally in the ataxic paraplegia of Gowers.

Rosen's Method for Nævus.—Thread through the growth silk previously soaked in alcohol and perchloride of iron, and leave it in for a week.

Roser's Position.—Keeping the head dependent over the end of the operating-table. Adopted in Kocher's operation, and tracheotomy.

Rosin's Test for Bile.—To the suspected fluid contained in an inclined test tube, add a few drops of tincture of iodine. In the presence of bile a grass-green ring is formed at the point of contact.

Rotterinum.—An antiseptic and disinfectant mixture, which is officially recommended in Southern Germany in lieu of bichloride and carbolic acid. It is composed of:

```
℞ Acidi citrici,
    Thymoli.....................................āā gr. iss
    Acidi salicylici ............................. gr. x
    Acidi boracici........  ..................... gr. xlv.
    Zinci chloridi,
    Zinci sulphocarbolatis.....................āā gr lxxv
    M. et ft. pastilli No. iv.  S. One pastil to a quart of water
```

Salacetol (Acetol-salicylic-ester).—An antirheumatic and antiseptic.

Saligenin.—A compound recently obtained by Dr. Loderer and administered for gout, in which it is said to remove in a very short time the fever, swelling, and pain. The disagreeable effects of salicylic acid—digestive disturbances, cyanosis, tinnitus aurium—are said to be lacking, and its action is more powerful and of longer duration. The drug is very soluble in water, and the daily dose is 3 gm. (gr. xlv.) in divided doses.

Salinaphthol.—Another name for Betol.

Salipyrin.—This is the salicylate of antipyrin, containing fifty-eight per cent of antipyrin and forty-two per cent of salicylic acid. It is a white crystalline powder, with a faint agreeable odor and sweet taste, practically insoluble in water, but soluble in alcohol and ether. In doses of gr. xv., four or five times daily, it is very useful in cases in which antipyretic and analgesic effects are desired. Its best results are seen in influenza, although it is often very efficacious in muscular rheumatism and tonsillitis.

Salitanuol.—A condensation product of salicylic and gallic acids. It is a white, amorphous powder, insoluble in water and the usual solvents. It has been used as a surgical antiseptic powder.

Salocoll.—Another name for Phenocoll Salicylate.

Salophen (Acetyl-para-amydosalol).—A white, odorless, tasteless powder, almost insoluble in water, and often of service in acute articular rheumatism when salicylate of sodium has failed. It has little if any ill effects on the stomach, heart, or kidney, and may be given in doses of gr. xlv. to xc. a day. It is also of service in habitual headache without known cause, in facial neuralgia following influenza, in lumbago, and general muscular pains. Recent observations have shown that this drug is an efficient antipruritic agent. Drews has reported (*Therap. Monatshefte*, March, 1898) cases of prurigo, urticaria, diabetic pruritus, and eczema, in which the distressing itching was relieved by 15-grain doses every two hours. In chronic articular rheumatism he does not find it superior to other remedies.

Salosantal.—A combination of salol and sandal oil, of value in gonorrhœa and bladder infection. Best dispensed in soft capsule.

Salumin.—Another name for Aluminium Salicylate.
10

Salve Muslins.—Ointments spread by machinery on undressed muslin or mull. The basis is benzoated mutton tallow and wax.

Sanarelli's Serum.—This is a serum used for yellow fever. Sanarelli states that it is bactericidal, but not antitoxic, and is efficacious in yellow fever only at an early stage and before the occurrence of serious organic changes in the various viscera as the result of excessive toxins.

Sanatogen.—Said to be a combination of sodium and casein glycero-phosphate. It contains about thirteen per cent of nitrogen, and is used as a nutrient tonic, in doses of a teaspoonful with meals.

Sänger's Gonorrhœal Maculæ. — These consist of small, red papules, situated about the openings of the Bartholinian glands. They are, as a rule, indications of chronic gonorrhœa.

Sanguinoform.—A preparation obtained by Wartenberg from the "embryonic blood-forming organs." It is a dry powder, with a pleasant taste. Advocated in anæmia, in doses of half a teaspoonful for an adult, three times a day. For a child, in proportion to the age.

Sanmetto.—Oil of sandalwood and fluid extract of saw palmetto. Used in genito-urinary diseases. Dose, ℨi. four times a day.

Sanoform.—Obtained by the action of iodine on methyl salicylate. A white, crystalline powder, used as a substitute for iodoform.

Sanose. — An albumen preparation, said to contain eighty per cent albumose. It is a white, odorless, and tasteless powder, which forms an emulsion when stirred with water.

Saprol.—A product of petroleum refining. Used as a disinfectant.

Savonal.—A medicated soap, introduced by Müller and Grube. It is made by mixing in the cold, olive oil, potash lye, and alcohol until full saponification, adding dilute hydrochloric acid, and then the alkaline mother liquid, until perfect neutralization. Evaporation to salve consistency.

Scarlet Fever : New Clinical Sign. — P. Meyer (*Presse Méd.*, March 5, 1898) has observed numbness of both hands combined with formication. Occasionally the sign exists in the feet alone, or in combination with that of the hands. It occurs in the eruptive stage, lasts from a few minutes to several days, and is usually intermittent.

Schenk Theory.—A theory set forth by Schenk, of Vienna, by which he claims that it is possible to govern the process of gestation so as to determine the sex of human offspring. It is claimed that evolution depends largely upon the amount of sugar supplied to the growing organism. If a male child is desired the mother must absolutely deny herself sweets. As the author himself puts the matter: "When no sugar is secreted, not even the smallest quantity, then the ovum will be developed which is qualified to become a male individual."

Schleich Method (General Anæsthesia). — After experimentation upon animals, the originator settled upon the following three mixtures as offering the greatest number of points of advantage over older methods. While in the inventor's hands the method has been used in hundreds of cases with unfailing success, some recent investigators have found in it much to condemn.

Mixture I.—(Boiling point, 38° C.)
    Chloroform .... 45 parts.
    Petroleum ether................................. 15 "
    Sulph. ether .... 180 "

Mixture II.—(Boiling point, 40° C.)
    Chloroform....................................... 45 parts.
    Petroleum ether................................. 15 "
    Sulph. ether .... 150 "

Mixture III.—(Boiling point, 42° C.)
    Chloroform .... 30 parts.
    Petroleum ether................................. 15 "
    Sulph. ether..................................... 80 "

Small doses are urged. Of Mixture I., 30 gm. for an operation lasting twenty minutes.

Schleich Method (Local).—A method of local anæsthesia in which three strengths of so-called sal anæstheticum are employed, as in the following table:

	I. Strong.	II. Normal.	III. Weak.
Cocain. hydrochlor	0.2	0.1	0.01
Morphin. hydrochlor	0.025	0.025	0.005
Natr. chlorat. steril	0.2	0.2	0.2

Each of these combinations is to be dissolved in 100 c.c. of water. Tablets of the above strength are to be had, and are convenient for extemporaneous solutions.

Schlunge's Sign of Intestinal Obstruction.—Total absence of peristalsis at any point below the seat of strangulation, and dilatation of the intestines above.

Schönlein's Disease (Peliosis Rheumatica). — This disease, also known as purpura rheumatica, consists of a purpuric skin eruption combined with rheumatic pains in the various joints. With these there may be an elevation of temperature, diminution of appetite, and general weakness. The mucous membranes are not involved, and recurrences are not infrequent.

Schott Method.—A system of resistance gymnastics in which the patient makes slow, systematic movements, which are opposed by the operator. The breathing must be kept quiet. The resistance is made by the hand of the operator held flat. Strict rules have been laid down for carrying out the various movements. It is advocated in circulatory disturbances, all enfeebled conditions of the heart, particularly when there is lack of compensation. Contraindications are advanced arterial sclerosis and grave complications. A system of baths, with or without massage, forms a part of the treatment. Mitral cases, particularly those of regurgitation, and all cases of slight incompensation, are mostly benefited. It is also useful in anæmia and chronic rheumatism.

This method is largely employed in Nauheim, Germany, where it was originally introduced. Nauheim bath salts in compact tablets for home treatment are now on the market.

Sclerogenic Method.—Lannelongue's method of injecting chloride of zinc for hernia has been extended by Coudray to tuberculosis of the larger joints, congenital dislocation of the hip, inoperable tumors, etc.

Scoliocoiditis.—A name introduced and used by Nothnagel to designate appendicitis.

Scopolamine.—Isomeric with cocaine, having an ac-

tion like that of hyoscyamine.   In paroxysmal excitement,
¼ to 1 mgm. hypodermatically.   Useful in acute mania.

Seiler's Antiseptic Tablets.—Used quite extensively
for nasal catarrh.   They consist of:

> Sodii bicarb. et sodii bibor.,
> Sodii benzoat. et sodii salicyl.
> Eucalyptol et thymol,
> Menthol et ol. gaultheriæ,

in varying proportions.

Sequardin.—Testicular extract.

Serum, Anti-Amaryl.—A term applied by Sanarelli to
his yellow-fever antitoxin.

Serum Therapy.—Serums have been prepared and
used with varying success in diphtheria, tetanus, tuberculosis, streptococcic and staphylococcic infections, typhoid
fever, bubonic plague, rabies, cholera, smallpox, yellow
fever, pneumonia, anthrax, syphilis, snake poisoning,
typhus fever, and cancer.

Sideroscope (Asmus).—An instrument used to detect
fragments of steel and iron within the eye and beneath the
skin.   The apparatus is described in a little book entitled
"Das Sideroskop und seine Anwendung," published by
Bergmann, of Wiesbaden.

Silver Citrate (Also called Itrol and Credé's Antiseptic).—This is a soluble form of metallic silver, and consists of a fine, white, odorless powder.   It stains as does
nitrate of silver, but the stain may be removed with water
or ether.   It has been used in the form of powder for the
antiseptic treatment of wounds, in solutions of 1:4,000
or 5,000 for disinfecting wounds, and in strength of

1:10,000 as a gargle. The ointment is said to have given favorable results in various septic conditions, such as lymphangitis, phlegmons, and the septic complications of scarlet fever and diphtheria.

Silver Lactate (Also known as Actol).—Another soluble preparation of silver, with highly antiseptic properties and great power of penetrating tissues. In cases of infectious inflammations, it has been used in the form of subcutaneous injections of from $\frac{3}{4}$ to 3 grains. Larger doses give rise to aseptic destruction of tissue.

Sitophobia.—A symptom of insanity, consisting of absolute refusal to take food.

Skiagraphy.—The most approved term for Roentgenray photography.

Skoda's Resonance.—A tympanitic percussion note heard in the infraclavicular region, in pleurisy with effusion.

Sleeping-Sickness.—A disease occurring among the negroes, chiefly on the west coast of Africa, characterized by a lethargic state lasting for weeks or months, and terminating in death. The filaria perstans seems to be a causative factor. See "Negro Lethargy."

Sodium Cinnamate.—Recommended, as is also cinnamic acid, in the treatment of pulmonary tuberculosis. It is a crystalline, white powder, soluble in hot water, alcohol, and fatty bodies. It has been recommended by Landrer as an intravenous injection in water or in physiological salt solution, after sterilization. A vein at the bend of the elbow is chosen and made prominent by compression.

Dose, gr. $\frac{1}{64}$ every forty-eight hours, increased by from gr. $\frac{1}{32}$ to $\frac{1}{64}$ up to a maximum dose of gr. $\frac{1}{8}$. Treatment should extend over four to six months, after which it should be omitted for from one to two months.

Sodium Copaibate.—Recommended for gonorrhœa, in doses of gr. x. to xx. four to six times a day. It seems to be less irritating to the gastro-intestinal and genito-urinary apparatus than oil of copaiba.

Sodium Phosphate.—Advocated by Wolff in daily doses of ℨ iii. to iv., for urticaria.

Solutol.—Cresylic acid made soluble with sodium cresylate. Used as a disinfectant.

Solveol.—A concentrated solution of cresylic acid with sodium creosote. An internal antiseptic.

Somatose. — Albumose of meat freed from peptones, containing eighty-eight per cent of albumose and twelve per cent of peptone. A light yellow, granular powder, odorless, tasteless, and soluble in all the ordinary fluids. In conditions of debility from whatever cause, in convalescence, dyspepsia, and wherever an easily assimilable and invigorating food is required, somatose has proven very efficacious and has acted as a restorative tonic. The adult dose is gr. cl. to ℥ ss.; the dose for children, gr. l. to ℨ ii. Drews and Joachim state that it increases the quantity and quality of the milk in nursing women. It is best administered in milk, gruels, coffee, cocoa, and water, shortly before meals. Somatose chocolate, somatose cocoa, and somatose biscuit have recently been placed on the market.

Somnal.—Hypnotic of unknown composition. said to be composed of chloral and urethan in alcohol. Used in doses of ʒ ss.

Sozoiodol.—A mixture of sulphur, carbolic acid, and iodine, in the proportion of 7, 20, and 54. An antiseptic powder used as a substitute for iodoform.

Spectrotherapy.—A thérapeutic measure suggested by Apéry, through the employment of prismatically decomposed rays. The subject has not yet been investigated.

Spermatocystitis. —An inflammation of the seminal vesicles, often of blennorrhagic origin; the inflamed vesicles are painful on rectal pressure and at times during defecation. The method by expression, first employed by Alexander for diagnostic purposes, has subsequently been elaborated by Fuller as a therapeutic measure. The condition may be simply catarrhal or suppurative.

Spermin.—From the testicular fluid of bulls, guinea. pigs, etc., combined with glycerin; given hypodermatically in thirty-minim doses or half the quantity, for impotence, tabes dorsalis, neurasthenia, etc. It is said to regulate oxidation and metabolism of the organism, thus relieving the nervous system of strain. Senator, Ewald, Mendel, and Fürbringer have spoken highly in its favor.

Sphacelotoxin.—The active principle of ergot. Dose, gr. ss. to iss. Uses, the same as those of ergot.

Spleen Extract.—This has been used in the form of a fluid extract or emulsion. It is said to aid nutrition and digestion, to increase the cutaneous circulation, the hæmoglobin, and red blood cells. The pulse becomes accelerated and the temperature slightly elevated. The dose of the fluid extract is three to four tablespoonfuls daily; that of the emulsion, ℨ i. four times a day. It may also be administered in tablet form.

Staphyloangina.—A term suggested by J. E. Walsh for pseudo-membranous inflammations of the throat, dependent upon staphylococci.

Staphylococcia. — A generalized infection with the staphylococcus. Usually local manifestations (furuncle, carbuncle, phlegmon) are present.

Stauungspapilla (Choked Disc). — Most frequently observed in cerebral tumors, tabes dorsalis, and paresis. May be secondary to retinitis.

Stenocardia.—Angina Pectoris.

Steresol.—A brown fluid, containing tolu, phenol, various gums, and alcohol. Used as a varnish, like collodion.

Sterisol.—A compound of formaldehyde, sodium chloride, and potassium phosphate. Used as an internal antiseptic.

Stethonoscope.—A newly patented instrument for auscultation, constructed on the principle of the microphone. It is said by the English manufacturers to be superior to the phonendoscope. It may be attached to a binaural stethoscope.

Stoker's Cramp.—Violent cramps in the muscles of the legs, back, and about the chest, affecting men (usually dissipated) who work as stokers in steamships. There may be unconsciousness, lasting from five minutes to two hours. Abdominal pain and constipation follow. One attack seems to predispose to another. Treatment: Large doses of opium, followed by castor oil and Epsom salts.— R. M. MYERS.

Streptoangina.—Used by J. E. Walsh to designate a pseudo-membranous deposit in the throat, due to streptococci.

Strobila.—A complete tapeworm.

Strontium Bromide.—Recommended in painful dyspepsias and in affections in which bromide of potassium is useful, not producing the ill effects of the latter drug. In epilepsy, hysteria, and cephalalgia it is said to be of value. Dose, ℨ ss. to i. per day.

Strontium Iodide.—A whitish-yellow powder, with a bitter taste, soluble in water and alcohol. Used as a substitute for potassium iodide in rheumatism, asthma, etc. Dose, gr. v. to xv.

Strontium Lactate.—White powder, soluble in alcohol and water. Used as an anthelmintic and tonic. Dose, gr. v. to xx. It is said to diminish albumin in the urine, in daily doses of ℨ ss. to i.

Strontium Salicylate. — Antirheumatic and general tonic, in doses of gr. x. to ℨ ss.

Strychnine Nitrate.—Useful in the treatment of alcoholism.

℞ Strychninæ nitratis.................. .... gr. $\frac{1}{15}$
  Atropinæ sulphatis .... gr. $\frac{1}{300}$
  Aquæ destillatæ........................... .... ℔ x.
S. Inject three times daily.

On the following day decrease to gr. $\frac{1}{20}$ with gr $\frac{1}{200}$ of atropine.

Stryone.—

> ℞ Liquid storax,
> Balsam of Peru...............................āā gr. iv.
> Distilled water,
> Alcohol.......... ....āā ℥ iiss.

Used in suppurative otitis media by instillation. Recommended by Bolt.

Styphage of Bailly.—A revulsive, by means of pledgets of cotton wet with chloride of methyl. Used along the course of painful nerve trunks, as in neuralgia, sciatica, etc.

Stypticin.—The hydrochloric-acid compound of cotarnine, one of the opium alkaloids, forming long, silky crystals. Used in uterine hemorrhages, in doses of gr. i. by the mouth, or half the dose hypodermatically.

Styra-Phenol.—A dressing for wounds and skin diseases, containing no grease.

Sulphocarbol.—See Aseptol.

Sulphonal (Diethyl-sulfone-methyl-methane).—A colorless, tasteless, odorless powder, rather insoluble in cold water, but soluble in warm water, hot coffee, soups, or alcohol. It is a prompt, efficient, and safe hypnotic in doses of gr. xv. to xxx., given before bedtime. In insomnias and neuroses free from pain, it is very useful. It requires several hours before its effects are produced. It is said to arrest the night sweats of phthisis. No ill effects are produced on the heart, respiration, or stomach. If given one to two hours before the effect is desired, it produces a sleep of six to eight hours. This drug is not a narcotic, but a hypnotic, and is not followed by a habit from continued use. Its continuous administration may, however, produce damaging effects and give rise to sulphonal poisoning, with physical and mental weakness, anorexia, vomiting, and red urine, due to the presence of hæmatoporphyrin. In three days, if the drug be discontinued, it will be entirely eliminated.

Suprarenal Extract.—From the suprarenal glands of oxen and sheep. Indicated in conditions associated with loss of muscular powers, asthenia, neurasthenia, Addison's disease, pernicious anæmia, and diabetes mellitus. The saccharated extract is best used in aqueous solution. An astringent action of decided power is obtained from a two- to four-per-cent solution; the ischæmia obtained persists longer than from other known astringents. H. L. Swain recommends local applications of an aqueous solution for hay fever and various congestive and hypertrophic processes of the nasal membranes. Upon the conjunctiva it is used without irritation or secondary congestion, and is highly esteemed as an adjuvant to cocaine. Also recommended by W. H. Bates in congestion of the mucous membranes of the ear, nose, and in urethral stricture. He says it has proven efficacious in angina pectoris. The best method of administration for acting on the heart and other internal organs is to place it carefully on the tongue, where it is absorbed by the mucous membrane.

Surmay's Operation.—Also known as jejunostomy. The first portion of the jejunum is drawn forward and sutured laterally to the anterior wall of the abdomen; in a few days a fistulous opening is made with a Paquelin cautery. Feeding through this opening can be carried out without difficulty. This operation is mostly applicable to neoplasms of the stomach which are too extensive to permit of resection or anastomosis.

Svapnia.—Said to contain the anodyne and soporific alkaloids—codeine, narceine, and morphine—excluding the convulsive and more poisonous ones. It has a standard strength of ten per cent of morphine.

Symphorol. — Caffeine-sulphonic acid. Said to be a reliable diuretic. Dose, gr. xv. four times daily.

Syndactylism,—The union of two or more fingers.

Syphilomania.—A tendency on the part of those cured of syphilis, or some venereal disease, to attribute every subsequent excoriation, herpes, or accidental lesion to the original syphilitic infection.

Syphilophobia.—A morbid fear of syphilis on the part of those possibly infected with other venereal disease or who have been exposed to infection.

Tallerman-Sheffield hot dry-air apparatus, used in the Tallerman treatment.

Tallerman Treatment.—This method consists in the local application of superheated dry air, by the introduction of the part into a cylinder or inclosed space. Used in the various forms of rheumatism and gout, sciatica, lumbago, sprains, synovitis, painful joints, rheumatoid arthritis, and gonorrhœal rheumatism. Séance from one-half to one hour. Although the Tallerman apparatus is protected by patent, it is said by the inventor that no case is permitted to be treated without direct medical approval and supervision. The limb is protected by lint and placed in the cylinder; the required temperature is quickly obtained, and the time allowed is from forty-five to sixty minutes. At intervals of twenty minutes the door is opened and the limb dried if the patient perspires freely. The degree of temperature varies with the case and with the patient, ranging from 250° to 400° F. The baths are given generally every second day. After the bath an hour is allowed for "cooling off."

Tannalbin (Albuminate of Tannin).—A yellowish powder, containing fifty per cent of tannin, and used in chronic diarrhœas, especially those of tuberculous origin. Adult dose, gr. xv.; for children below the age of four years, gr. vi. three or four times daily, or p.r.n.

Tannigen.—A white, odorless, tasteless powder, made from tannic and acetic acid. Its great advantage over tannic acid is that it passes the mucous membrane of the stomach unchanged, and thus acts as a true intestinal antiseptic. In doses of from gr. iii. to xv. it has been used in chronic diarrhœa, chronic dysentery, summer diarrhœa of children, and in all forms of diarrhœa of non-infectious origin.

Tannoform.—A reddish-white powder, recommended in pruritus vulvæ of diabetics and in hyperidrosis. Used either pure or mixed with starch.

Tannon.—Same as Tannopin.

Tannopin. — A brownish, tasteless, slightly hygroscopic powder, insoluble in water, alcohol, and ether, but dissolving slowly in weak alkalies. It is an intestinal astringent, which is a compound of tannin (eighty-seven per cent) with hexamethylentetramin or urotropin (thirteen per cent). It is said to pass the stomach unaltered, and to liberate its astringent constituents in the intestinal canal. It has been used in doses of gr. iii. to x., three times daily, for acute and chronic intestinal catarrh, the diarrhœa of typhoid, and tuberculosis. This drug was originally introduced by Nicolaier, in the treatment of affections of the urinary apparatus.

Tannosal.—A combination of tannic acid and creosote. It is a brown powder, soluble in water.

Tardieu's Spots. — Ecchymotic stains on the lungs, observed after death from asphyxia.

Telelectro-Therapeutics.—Gerest, of Lyons, is said to cure hysteric paralysis by a series of electric discharges in the neighborhood of the patient, without contact. The effect is evidently similar to that of suggestion.

Telegony.—The transmission, to the children of a second husband, of characteristics possessed by the first.

Telepathy has been dignified as a branch of science by the presentation of the subject to the British Medical Association by Professor Crookes. Mr. Meacham's theory to account for its phenomena in medicine includes "a thought-vibrating, circumambient ether," which transmits ideas through a "universal psychic menstruum" to the little neurons which project their tufts in hypersensitive search for and response to countless vibrations.

Tellurate of Potash.—An antisudorific. Dose, gr. $\frac{1}{8}$ to gr. i. per day in pill form.

Tellurate of Soda.—Dose, gr. ⅓ daily in pill form, or

R Sodæ tellurat................................... gr. ij,
    Alcohol.... .... ℥ ij.
M. S. Teaspoonful morning and night in sweetened water.

Eight to ten days suffice for a cure of decided hyperidrosis (Barth).

Terebene.—An agreeable preparation of turpentine, made from the latter by the gradual addition of sulphuric acid. It is a pale yellow fluid with an odor of pine wood, slightly soluble in water, but soluble in alcohol and ether. It is used as a stimulant expectorant, antiseptic, and deodorizer. Very serviceable in subacute and chronic bronchitis, winter cough, the flatulence of fermentation, and occasionally in sloughing wounds. In phthisical affections it may be given by inhalation. The dose is ℩ iii. to x., which may be given in the form of capsules or emulsion. Its advantages over turpentine are its more palatable qualities and the lessened danger of renal injury.

Terpin Hydrate.—A substitute for turpentine. It consists of crystals which are insoluble in cold water, but soluble in chloroform, alcohol, and ether. In small doses it increases the secretion of the bronchial mucous membrane and also that of the kidney. It is used in chronic bronchitis and sometimes in nephritis on account of its diuretic and diaphoretic properties. The dose is gr. ii. to viii., best given in pill form or alcohol.

Terpinol.—A derivative of terpin. It diminishes excessive bronchial secretions in daily dose of gr. v. to xv.

Terralin.—A mixture of calcined magnesia, kaolin, silica, lanolin, glycerin, and some antiseptic. Introduced
11

by Tschkoff. Used as a vehicle which does not readily alter on keeping and can be easily removed from the skin by water.

Terrol.—A fluid petroleum introduced as a substitute for cod-liver oil. Dose, ℥ ii.

Testaden.—Dried testicular extract. Dose, 30 grains t.i.d. for neurasthenia, impotence, and spinal diseases.

Tetanus Antitoxin.—A prophylactic and at times a curative agent for idiopathic or traumatic tetanus. It has lately been used in the form of intra-cranial injections at the General Hospital in Passaic, N. J., and with a successful result. It is dispensed in vials of 10 c.c. Roux and Borel state that a few drops of antitoxin injected intra-cerebrally produce the same effect as large quantities by hypodermatic injection or directly into the blood.

The extract of the brain of warm-blooded animals (sheep, rabbit) has recently been employed as an antitetanic injection, with apparently good results.

Tetronal.—Closely allied to sulphonal and trional, but not so efficacious as either. It is less soluble than trional, and more toxic in its action. Dose, gr. iv. to viii. It is rarely used in this country.

Thermodin (Acetyl-π-ethoxy-phenyl-urethane.) — An antiseptic and antipyretic. Dose, gr. v.-x.

Thermomassage.—A combination of heat application and massage by means of smoothing-iron-like and rolling instruments filled with acetate of sodium. Advocated by Goldscheider (*Zeitschrift f. diät. u. Physik. Therapie*, Bd. 1,

H. 3, 1898).   Useful in rheumatism, neuralgia, and other affections.

Thermopsychrophor.—An instrument devised by Dr. Scharff, of Stettin, for the treatment of chronic prostatitis. The instrument has for its object the rapid application of alternating heat and cold (*Lancet*, November 12, 1898).

Thermotherapy.—The application of heat has recently been extended in many directions. Hot spray baths in lichen planus, psoriasis, itching affections; hot-water injections in gonorrhœa; hot coil in chancroid (40° to 45° C.); poultice kept at a given temperature by a metallic cap fitting over it and through whose summit a supply tube passes (Quincke); hot bag to leg ulcers for many hours together; hot vapor in obstinate chronic eczema. Dry heat to chronically inflamed joints, by means of special apparatus (see Tallerman method, page 158). The application of superheated air seems susceptible of wide application not only in lithæmic conditions, but in respiratory affections, tonsillitis, conjunctivitis, etc. The apparatus in use must be capable of furnishing a temperature of from 300° to 350° F.

Thermal baths have been found efficacious in tabes dorsalis.

Thiersch's Solution.—

Salicylic acid	2
Boric acid	12
Water	1,000

A valuable antiseptic wash for nose, throat, stomach, and bladder.

Thilinin.—Obtained by the action of sulphur on lanolin so that the latter contains three per cent of the former. It has the appearance of vaseline and is used in eczema, etc.

Thiocol.—The potassium salt of guaiacol-sulphonic acid, containing about sixty per cent of guaiacol. It is a fine white powder, with a bitter-sweet taste, soluble in water, recommended as a substitute for guaiacol in phthisis, chronic bronchitis, and scrofula. Dose gr. viii.-xx. t.i.d. Best administered in solution.

Thioform.—A yellowish-brown, odorless powder, occasionally used to supplant iodoform on account of its odorless and non-toxic properties.

Thiol.—A substitute for ichthyol, obtained by the action of sulphur on the oil of gas of commerce. It occurs either as a black powder which is soluble in water or as a black, syrupy liquid. As a vaginal dressing, tampons are used with thiol in ten- to twenty-per-cent solution in glycerin. For application to the skin it may be used in five- to ten-per-cent ointment, or in burns the liquid thiol may be used in its purity or with equal parts of water. It has an advantage over ichthyol in being odorless.

Thionin (Hoyer).—A basic dye used for staining blood specimens and the plasmodium malariæ. A saturated solution in thirty-three-per-cent alcohol is usually employed.

Thioresorcin.—A combination of resorcin and sulphur, consisting of a yellowish-white, odorless powder insoluble in water, but slightly soluble in alcohol and ether. Used as a substitute for iodoform and in ointment form (five to ten per cent) as a remedy for psoriasis, seborrhœal eczema, and other skin lesions.

Thiosavonal.—A soluble sulphur soap made by saponi‑fying Riedel's sulphur oil with potash lye and adding empyreumatic oil of birch.

Thiosinamin.—A chemical product obtained by reaction with ammonia from the volatile oil of mustard. Occurs as white crystals with a faint garlic odor and of slightly bitter taste. This has been used in enlargement of glands, carcinoma, lupus, adenoid, keloid, and the absorption of cicatrices. It is soluble in water, alcohol, and glycerin. Watery solutions are less permanent. Dose, ℔ vii. to x. of a fifteen-per-cent solution given hypodermatically. It may be given in doses of gr. ⅔ to gr. ii. for painful gastric tumors.

Thomsen's Disease.—See Myotonia Congenita, page 116.

Thymacetin.—A white powder slightly soluble in water. Dose, gr. iv. to xii. as an analgesic.

Thymol Biniodide.—See Aristol.

Thymus Gland.—Preparations of this gland have been used internally on the same principles which govern the use of thyroid. It is supposed to benefit certain systemic disturbances arising in persons in whom the thymus gland became atrophied at too early a date. Some have used it instead of the thyroid. Its main application is in the treatment of simple goitre. The dose is half an ounce several times a week. It is best administered in capsules or tablets.

Thyol.—See Thiol.

Thyreoid.—Name used by P. D. & Co. for a concentrated thyroid extract.

Thyrocol.—A proprietary preparation used in psoriasis; given in palatinoids.

Thyroglandin.—An English thyroid preparation said to be four times as strong as the fresh gland.

Thyroid.—A substance prepared from the fresh gland of sheep and used during the past few years in a variety of conditions with varying though usually successful results. The conditions in which it has proved most efficacious are myxœdema, cretinism, lipomatosis universalis, cachexia strumipriva, insanity, and Basedow's disease. It has been recommended by Hertoghe as a galactogogue. Spastic torticollis, tetany, acromegaly, Parkinson's disease, hæmophilia, enlarged prostate, fibroid tumors of the uterus, psoriasis, and premature grayness have all in turn been subjected to its use. In an instance recorded by Mossé an infant at the breast having a marked bilobed goitre was cured by the administration of the drug to the mother. The hypodermatic administration of the liquid extract and the grafting of the fresh gland have become obsolete, and the tendency of to-day is to use the powder, capsules, or tablet prepared from the desiccated fresh gland. It is advisable to begin with small doses, gradually increased until the desired effect is reached. Beginning with one or two grains a day, we may administer as much as fifteen grains daily, always being on our guard for poisonous effects. Especial attention should be given to the respiratory and cardiac apparatus, and at the first appearance of rapid pulse, embarrassed respiration, rise of temperature, vertigo, irritability, and gastric disturbance its use should be abandoned.

Tic Rotatoire.—The same as *tic giratoire*. A clonic spasm of the obliquus capitis inferior. According to Meyer (*Wiener klinische Rundschau*, January, 1898), tic rotatoire depends in the majority of cases upon functional disturbances of the brain, and is caused by mental or physical excitement. In rare cases it depends upon organic changes in the brain, or may be due to pathological changes in the course of the spinal accessory nerve itself, or in its branches supplying the sterno-cleido-mastoid and trapezius.

Toison's Solution.—Used with Thoma-Zeiss apparatus for counting the red and white blood cells with a single solution.

A.

Methyl violet	0.025
Neutral glycerin	30.0
Distilled water	80.0

B.

Sodium chloride	1.0
Sulphate of sodium	8.0
Distilled water	80.0

Mix A and B, and filter. Use the red-blood-cell pipette for the counting. With this fluid the white blood cells appear violet, the red blood cells yellow-green. Useful in diseases associated with marked leucocytosis.

Toluidin-Blue.—Used as a collyrium in inflammatory eye affections in solution of 1:1,000. It seems materially to lessen pus formation. Water readily removes the stain from the skin.

Toluol.—A preparation recommended by Loeffler for the local treatment of diphtheria.

℞ Solution of chloride of iron	2 c.c.
Menthol	5 "
Alcohol, absolute	30 "
Toluene	18 "

Toxicodendrol.—A non-volatile oil discovered in 1895 by Pfaff, of Harvard, and said by him to be the etiological factor in the production of dermatitis venenata or ivy poisoning. It is soluble in oil but insoluble in water, and for this reason salves and oily applications, it is thought, only tend to spread the affection, while washing with soap and water or alcohol removes the poison. Pfaff says that the oil is very readily destroyed by an alcoholic solution of lead acetate.

Trapp's Coefficient.—Used for the quantitative determination of the total amount of solids excreted in the urine. The figure employed is 2, and the method of procedure is the same as in Haeser's coefficient.

Traube's Semilunar Space.—This is a region bounded by the liver, lung, spleen, and free border of the ribs. Under normal conditions the percussion note in this locality is tympanitic. In left-sided pleurisy with effusion the note becomes flat, thus forming a differential aid in diagnosticating this condition from pneumonia.

Traumaticin.—A thick liquid similar in appearance to collodion and consisting of a saturated or 1:10 solution of guttapercha in chloroform. It is used as a vehicle for chrysarobin, ichthyol, anthrarobin, aristol, etc., forming a thin pellicle when the chloroform evaporates.

Triacid.—This is the triple neutral stain of Ehrlich used so extensively for blood specimens. The constituents are acid fuchsin (extra), methyl green, and orange G mixed in certain proportions. Thus:

Saturated watery solution of acid fuchsin	50
" " " " orange G.................	70
" " " " methyl green	80
Distilled water......................................	150
Absolute alcohol.....................................	80
Glycerin..	20

The specimen being fixed is stained at once in from five to ten minutes. The stain is called neutral, because the ε granulations, otherwise known as neutrophile granulations, take up this stain, thus differing from the eosin hæmatoxylon stain, which does not affect the neutrophile granulations. With this solution the red blood cells appear redviolet, the nuclei of the leucocytes light green or blue-gray, the nuclei of the erythroblasts (only present under abnormal conditions, or see page 57) dark blue or dark green, the α or eosinophile granulations red, the ε or neutrophile granulations violet. This solution is also used to determine the presence or absence of fibrin in the sputum. A sputum rich in fibrin (pneumonia) is stained red, whereas a sputum with an abundance of mucus and leucocytes (bronchitis) is stained blue-green. This test is both microand macroscopical.

Trichloracetic Acid.—Consists of large deliquescent crystals soluble in water and alcohol. Used for the removal of corns, warts, callosities, hypertrophies of the nasal cavities and pharynx. As it is very caustic and spreading in its action, it must be limited to the seat of the lesion. Needless to say, a previous application of cocaine is requisite when using it on mucous membranes.

Trichotillomania.—The habit of pulling out hairs from the head, brows, or beard. It exists also in mammals and birds, especially parrots show the same habit in a continual plucking out of feathers.

Trigger Finger is an annoying affection characterized by the sudden and involuntary locking of the finger when it is flexed or extended to a certain point. The finger cannot be flexed or extended without a powerful effort, or without the aid of the other hand. Sometimes this effort is attended by a distinct snap. Thought by Jeannin to be due to a narrowing of the tendon sheath of the flexors. At times the affection is self-limited. It is also known as lock-finger, snap-finger, jerk-finger, spring-finger, and digitus recellens. It affects one digit as a rule, and the middle one most frequently. Females are more frequently attacked than males. Some cases are associated with gout and rheumatism, others are due to occupation—musicians, seamstresses, etc.

Treatment: Widening of the sheath and complete removal of the obstacle.

Trikresol.—A clear, watery fluid said to have three times the disinfectant value of carbolic acid, recommended by Dr. E. A. de Schweinitz, of Washington, as an antiseptic for the preservation of collyria. An aqueous solution (1:1000 to 1:500) is recommended as a vehicle for cocaine, atropine, eserine, etc. It is said to produce no burning, irritation, nor pain.

It has recently been recommended by MacGowan, of California (*Journ. of Cut. and Gen.-Ur. Dis.*, May, 1899), for alopecia areata, and applied in its pure state for trichophytosis of the scalp.

Trinitrin.—Same as nitroglycerin. Recently recommended for hæmoptysis and laryngismus stridulus.

Trional.—A white crystalline substance, of bitter taste, almost insoluble in water, but soluble in alcohol and ether. It acts as an hypnotic and sedative in doses of gr. xv. to xx., Its action and dose are midway between those of sulphonal and tetronal. Dose varies from gr. x. to xxx.; gr. xv. is a good average dose, and may be repeated in an hour. It is best administered in a cupful of warm milk, tea, sweetened water, or gruel.

Triphenin.—A coal-tar derivative. Antipyretic and analgesic. Dose, 5 to 10 gm.

Tritipalm.—A proprietary genito-urinary tonic, consisting of the fluid extract of fresh saw palmetto and triticin. Dose ℨi. t.i.d.

Tritol.—A name given by Dieterich to a diastasic extract of malt, twenty-five parts of which are sufficient to emulsify seventy-five parts of cod-liver or castor oil.

Tropocaine (Tropococaine).—A local anæsthetic which has been used in eye practice in lieu of cocaine. It was

first obtained from Javanese coca leaves, and is now procured synthetically by the decomposition of atropine and
hyoscyamine. It occurs in colorless crystals which are
very soluble in water. The strength of the solutions used
varies from two to six per cent. It is said to produce anæsthesia more quickly than cocaine, with less consequent
irritation and hyperæmia. Mydriasis rarely occurs and
accommodation is not affected. It differs from cocaine in
that its anæsthetic effects last longer, it does not reduce
swelling, and it is less dangerous.

Trousseau's Phenomenon of Tetany.—This consists
of an ability to reproduce the tonic spasm of the disease by
pressure on the large arteries and nerves of the arm. The
spasm continues so long as pressure is kept up.

Tubercle Bacilli.—Since the usual methods are not reliable in decomposed sputum, Housell advises to stain with
hot carbol fuchsin, and to place the specimen in a three-per-
cent solution of hydrochloric acid with absolute alcohol for
not less than ten minutes. Counter stain with methylene
blue in alcohol or corallin in absolute alcohol (Pappenheim).

Tuberculin (Koch).—At present this is mostly used as
a specific diagnostic agent in any stage of tuberculosis,
and especially in veterinary practice.

Tuberculin R.—A new tuberculin obtained by Koch
by means of a mechanical process for breaking the wax-like
enveloping membrane of tubercle bacilli so as to obtain the
constituents of the bacilli themselves. So far there have
been favorable reports in the treatment of lupus, and
various tuberculoses both general and local.

Tumenol.—Obtained by treating mineral oil with strong sulphuric acid. In oily solutions of ten to twenty-five per cent or in ointment, it allays cutaneous irritation and pruritus.

Tussol (Antipyrin Mandelate).—Used in pertussis.

Typewriter's Cramp.—One of the occupation neuroses. It has been suggested that a hammer be used for striking the keys, but, like the various devices for writer's cramp, all means are apt to fail except a change of occupation.

Tyree's Antiseptic Powder.—A proprietary preparation which is said to contain alum, biborate of sodium, eucalyptus, carbolic acid, thymol, wintergreen, and peppermint. It has been recommended for leucorrhœal and purulent discharges, intertrigo, eczema, etc.

Uffelmann's Reagent.—Used for the detection of lactic acid in the stomach contents. To 10 c.c. of a four-per-cent carbolic-acid solution add 20 c.c. of water and one or two drops of the tincture of chloride of iron. This gives an amethyst-blue solution. Add a few drops of the filtrate of the stomach contents, and if lactic acid is present the reagent assumes a green-yellow color, often spoken of as canary-yellow.

Unguentin.—This is said to be an alum and petrolatum ointment containing ichthyol (five per cent) with carbolic acid (two per cent).

Unit of Antitoxin.—This is ten times the amount of serum required to protect a guinea-pig weighing 250 grams when ten times the fatal dose of the toxin is mixed with the serum and injected subcutaneously.

Ural.—A mixture of chloral and urethrane, said to be a safer hypnotic than chloral. Dose, gr. xv. to xlv.

Uranium Nitrate.—A valuable addition to remedies for glycosuria. A large amount (gr. xc. daily) may be given without unpleasant symptoms. Best given in gradually increasing doses from gr. iii. twice daily after meals and in a large amount of water. Favorable reports have been made by Samuel West, C. Hubert Bond, and others.

Urea.—Used principally on account of its diuretic action and its ability to hold uric acid in solution. It produces its best effect in the ascites of biliary cirrhosis and cardiac disease, and in pleurisy with effusion. In renal affections it is less efficacious. In urinary lithiasis it may be given in ten-per-cent aqueous solution, of which a teaspoonful is to be taken every hour. Klemperer prescribes as follows:

```
℞ Ureæ puræ ............................................. 10
    Aquæ destillatæ  ................................... 200
M.  S. Tablespoonful every hour
```

This is increased to 15 in 200 in two days, and to 20 in 200 in two days more. Occasionally it causes a diarrhœa. The taste of the drug may be disguised with milk or seltzer. In uric-acid diathesis:

```
℞ Sodii bicarb.,
    Calc. carb.,
    Ureæ pur ...........................................āā 25
M.  S. Half a teaspoonful three or four times a day.
```

In pernicious malaria associated with coma, the hypodermatic administration of gr. xx. of bimuriate of urea and quinine has proven very efficacious.

Urea Salicylate (Ursal).—Used the same as salicylate of sodium, gr. vii. one to four times a day.

Urecidin.—A brownish granular substance prepared from lemon juice and citrate of lithia. Used in uric-acid diathesis, gout, urinary calculi, migraine, sexual neurasthenia. Dose, gr. xv. to 3 iss.

Urethane.—An ethylic ether of carbamic acid. This is a hypnotic in doses of gr. xv. to xlv., in solution; children, gr. v. to xv.

Urobilin.—This is reduced bilirubin or hydrobilirubin. It is found in the urine in diseases of the liver (atrophic and hypertrophic cirrhosis), fever, cerebral hemorrhage, infarcts, and in the hemorrhagic diatheses (scorbutus, morbus maculosus Werlhofii). It is also found in a particular form of icterus known as the urobilin icterus of Gerhardt.

TESTS: (*a*) To a few cubic centimetres of urine add ammonia in excess until the reaction is strongly alkaline, and filter; then add a few drops of a ten-per-cent chloride-of-zinc solution and filter again. Examine against a dark background, and the filtrate appears green, and with transmitted light rose-red. Inasmuch as the various urinary pigments may resemble this color reaction, it is safer to examine the solution with a spectroscope. With this means we see an absorption band between green and blue.

(*b*) *Grinbert's Test:* Take equal volumes of urine and hydrochloric acid and boil; then shake well with ether. In the presence of urobilin the ether appears a brownish-red and of a greenish fluorescence.

(*c*) *Rasmussen:* Take equal parts of urine and ether, add six or seven drops of tincture of iodine, and shake thoroughly. Set aside until the fluid separates into an upper layer of ether and iodine and a lower one of urine. If bile be present the lower layer turns green in the presence of biliverdin.

Uropherin.—A salicylate and benzoate have been prepared from theobromine lithium. Said to be more readily absorbed than diuretin and efficacious in smaller dose. It has a bitter taste. Dose, gr. xv.

Schmidt recommends for pleurisy and pericarditis of children:

℞ Uropherin salicylatis................................	5	gm.
Mucilaginis acaciæ.............................	15	"
Syr. simplicis.....................................	15	"
Vanillin...	0.001	"
Aquæ..	120.0	"

M.   S.   A dessertspoonful one to three times a day.

Urotropin (Hexamethylentetramin or Aminoform).—Belongs to the formaldehyde group. Obtained by a combination of formic aldehyde and ammonia. In half-gram doses it soon enters the urine and tends to produce a disinfectant and astringent effect. In pyelitis, chronic cystitis following prostatic hypertrophy and stricture of the urethra, and in the cystitis following exposure to cold ("Erkältungs-Cystitis" of the Germans) it works well and promptly, whereas in acute gonorrhœa, in gonorrhœal and tuberculous cystitis it works very poorly. It has been used in uric-acid diathesis and gout; gr. xv. to xx. in a glassful of water every morning before breakfast being recommended as a prophylactic.

Uvuloptosis.—A relaxed condition of the uvula and soft palate.

Vagabond Disease.—Pediculosis vestimentorum.

Validol (Menthol Valerianate or Valerianic-Acid Menthol).—An oily liquid, first introduced by Schwersenski, possessing the properties of dissolving considerable quantities of pure menthol, from which the irritating properties are also removed. It contains thirty per cent of menthol. Used in trigeminal neuralgia, cardialgia, catarrhal angina, neurasthenia, and hysteria. Dose, gtt. xii. on sugar p.r.n. It may be used externally for sore throat or as an inhalation for sea-sickness. Also used as a stomachic and carminative.

Valsol.—Vaselina oxygenata or vasogen.

Vasogen.—Vaseline treated by an excess of oxygen, which makes an emulsion with water. It forms an excipient for iodine, ichthyol, menthol, iodoform, chrysarobin, guaiacol, etc., which are incorporated during the manufacture. Its greatest advantage seems to be its ability to penetrate the pores of the skin more than any other remedy. The iodine vasogen is of service in epididymitis, inguinal swellings, and skin diseases. Vasogen mercurial ointment for the inunction treatment of syphilis can now be obtained in convenient though expensive capsules.

Vasothion.—A remedy said to contain ten per cent of sulphur with vasogen (*Apoth. Zeitung*, xiv., p. 155). It may be applied either pure, in ointment, or emulsion for chronic skin diseases.

Venesection and intravenous infusion of salt solution is a form of treatment in eclampsia advocated by Cutler (*Bost. Med. and Surg. Jour.*, March 30, 1899).

Vesuvin.—This staining agent is a triamido-azobenzene hydrochloride employed as a differential stain for the diphtheria bacillus.

12

Vicarious Urination.—A condition described by Dr. A. T. Rice, of Woodstock, Ont., in which, with almost total suppression of urine, there was an exudation of fluid from the anterior portions of the lower limbs between the knee and ankle. The fluid was voided three times daily at regular intervals, in gradually increasing amounts, the average being thirty to forty ounces a day. The fluid was amber-colored, with a specific gravity of 1.010, without albumin or sugar, and had a strong smell of urine on boiling, with a distinct ammoniacal smell after standing; uric acid was found on examination. There was no abrasion of the skin, no œdema; the fluid simply oozed out. For the want of a better term, the condition was named vicarious urination.

Vimbos.—A fluid beef, manufactured in Scotland, which contains, according to Dr. Stevenson Macadam:

Nitrogeneous organic matter	50.86
Fatty bodies	1.81
Saline matters	23.51
Moisture	24.12

Walker Gordon Milk, or "Modified Milk."—Depots have been established in various large cities to supply from the laboratory infant food with any desired proportion of fat, milk sugar, albuminoids, etc., according to the physician's prescription for each particular case, just as medicines are ordered from the druggist. Order blanks in prescription form are supplied on demand.

Weichselbaum's Diplococcus.—The diplococcus intracellularis of Weichselbaum is the recognized specific germ of epidemic cerebro-spinal meningitis. It has recently been demonstrated by Gwyn in the blood during life. It is found in the fluid obtained by lumbar puncture, and in the pus from joints.

Weil's Disease (Acute Febrile or Infectious Icterus).— A disease first described by Weil and characterized by the sudden invasion of chills, fever, jaundice, headache, pains in the back, limbs, and muscles, albuminuria, and enlargement of the spleen. Mild delirium and coma sometimes occur. Men are far more frequently attacked than women, and butchers seem to be particularly susceptible. Holz has reported an instance in a woman fifty-one years of age. The jaundice is an obstructive one, the stools being clay-colored and the urine rich in bile pigment. The disease lasts from one to two weeks as a rule, but recurrences are not infrequent. The treatment is purely symptomatic. Calomel or salts at the beginning; later antipyretics and stomachics. Later experiments seem to show that this disease is a proteus infection. Libman (*Phil. Med. Jour.*, March 18, 1899) prefers to designate these cases as "infections by the bacillus proteus (fluorescens)."

Westphal's Symptom (Loss of Knee-jerk).—Seen in locomotor ataxia, anterior poliomyelitis, myelitis in the lumbar region, multiple neuritis (alcohol, diphtheria, diabetes, trauma, etc.), general paresis, Friedreich's disease (hereditary ataxia), and occasionally in cachexia and pernicious anæmia.

Widal.—A name associated with the serum reaction, the serum test, or the serum diagnosis of typhoid fever.

The test is based upon the principle that if a drop of properly diluted blood serum, from a person suffering with typhoid fever, be brought into contact with a living culture of typhoid bacilli, the latter will lose their activity and become grouped or agglutinated into small masses or "clumps." This may be seen macroscopically or microscopically. For purposes of diagnosis, however, the latter method, with examination in the hanging drop, is the only reliable one. The dilution to be preferred is 1:10—*i.e.*, 1 part of serum to 10 of distilled water. Dilutions of a lesser degree are not so apt to be accurate. The culture used ought not to be more than twenty-four hours old. Any culture, however, in which the bacilli are active will answer. As a rule, the reaction is not found before the sixth or seventh day, though cases have been observed in which it was seen as early as the third. On the other hand, it may not appear until the relapse. From the statistics which have been gathered up to the present date, it may be said with all fairness that in ninety to ninety-five per cent of the cases a positive "Widal" means typhoid fever. The absence of this reaction does not, however, exclude typhoid. There is a margin of error in from five to ten per cent of the cases; this is accounted for by the fact that a positive Widal has been observed in cerebro-spinal meningitis, acute articular rheumatism, phthisis, pneumonia, influenza, meningitis. Recent experiments have shown that the blood of a person who has recovered from typhoid fever may give the Widal reaction for from six months to one year after the attack.

The blood for examination may be collected in capillary tubes or on cover glasses and slides. If collected on slides it is advisable to procure three drops, so that two may act as control tests for the third.

Williamson's Blood Test for Diabetes.—Place in a narrow test tube 40 c.mm. of water and 20 c.mm. of blood; to this add 1 c.c. of methyl-blue solution (1:6,000) and 40 c.mm. of liquor potassæ. Place the tube in a water-pot which is kept boiling. If the blood is that of a diabetic patient, the blue color disappears in four minutes and becomes yellow. In non-diabetic blood, the blue color remains.

Winckel's Disease.—An epidemic disease of the newborn, described by Winckel and characterized by cyanosis, afebrile icterus, hæmoglobinuria, and marked nervous symptoms. The course of the disease is rapid, death almost invariably resulting in from a few hours to four days.

Winkler's Bodies.—Spherical forms found in the lesions of lues. Each body has an eccentric light spot; they are stained with thiodin and toluidin-blue and discolored with formalin. Unna regards them as a product of syphilis differing from blastomycetes, hyaline products, etc.

Winkler's Reaction.—A test for determining the presence of free hydrochloric acid in the stomach contents. To a small quantity of stomach contents filtered into a porcelain dish or spoon add a few grains of dextrose; pour into this a few drops of α-naphthol solution and heat gently. The presence of free hydrochloric acid is shown by the development of a violet-blue color which soon changes to an inkish black.

Wolff's Mixture.—

Sodium sulphate	℥ i.
Potassium sulphate	ʒ i⅓.
Sodium chlorate	℥ i.
Sodium carbonate	ʒ vi.
Sodium borate	ʒ iiss.

M. S. Half a teaspoonful in a half glassful of water three times a day.

Used in cases of hyperchlorhydria or excessive production of hydrochloric acid.

Woillez's Disease.—An idiopathic pulmonary congestion, following an exposure or sometimes traumatism. It differs from pneumonia in the absence of severe and prolonged rigor. The temperature rises rapidly to 40° or 40.5° C., and there is slight expectoration. The affected side may appear increased in size; expiration is prolonged and inspiration somewhat interrupted. The duration is two to five days, and there may be a relapse. On postmortem there is found congestion without consolidation. Treatment is by dry-cupping, poultices to relieve the pain, and quinine internally.

Writer's Cramp.—According to Monell (*Med. Record*, July 23, 1898) this is not a neurosis, but the result of malnutrition. He advises first a " warming-up application " to quicken the circulation, then general nutritional muscular contractions, and finally refreshing, restful, nutritional applications; the total treatment requiring ten minutes.

Xanthopsia.—Yellow vision; sometimes observed after the administration of santonin.

Xeroform (Tribromphenol Bismuth).—An almost odorless and tasteless antiseptic powder for internal and external

medication. It has deodorizing qualities, diminishes secretions, and as a rule is non-irritating. It has been used
in intestinal affections (flatulence, constipation, atony),
tuberculous processes, and skin diseases. In powder and
salve form it has proved useful in diseases of the eye,
eczema of the lids, follicular and pustular conjunctivitis.
The internal dose is gr. viiss. t.i.d.

Ehrmann (*Wiener mediz. Blätter*, No. 22, 1898) has
found it very efficacious in balanitis, moist eczema, chan-
croids, suppurating buboes, and clean operative wounds.

Xerostomia.—Dry mouth.

Xylol.—A hydrocarbon used as an external application
in variola, and in microscopical work for the cleansing of
lenses and the dehydration of cut specimens. It resembles
benzene.

Yerbazin.—Used as a vehicle for quinine and other
bitter medicines.

Yerba Santa.—The syrup is used as a stimulant expectorant, but particularly as a vehicle for the administration
of quinine. It disguises the bitter taste of quinine better
than any other preparation. The aromatic syrup is preferred by some.

Ziehl's Solution.—A fluid used for the staining of tubercle and lepra bacilli. It consists of a five-per-cent aqueous solution of carbolic acid, to which is added one-tenth
its volume of a saturated alcoholic solution of fuchsin.
The specimen is first heated for three minutes in this solution, with the result that the entire specimen—tubercle
bacilli, epithelium, débris, other bacteria—is stained red.

Then the specimen is decolorized with a twenty- or thirty-per-cent solution of nitric acid. Owing to the tenacity with which the tubercle bacilli retain the solution, everything except these lose their color. The specimen may be examined with an oil immersion at once, or preferably after using a differential stain of methylene blue (five per cent) or Bismarck brown. The tubercle bacilli appear red, all other bacteria blue.

Zoolak.—A new name for Matzoon, applied to the preparation originally introduced into America by Dr. Dadirrian, to distinguish it from others.

Zoöphobia.—Emotional fear of animals as experienced by human beings.

ADDENDA.

Fehling's Solution.—In testing a specimen of urine for sugar with this solution care should be exercised lest the reduction be due to other substances, such as acetanilid, sulphonal, antipyrin, chloral hydrate, salol, salicylates, and senna. Before testing, always ask if the patient has taken any of these.

Ferripton.—Introduced by E. A. Kunze, of Radebeul, as a therapeutic agent for chlorosis and anæmia. It is said to contain four per cent of iron, seven per cent of proteids, and eighty-nine per cent of water.

Ferropyrin.—Chloride of iron and antipyrin, said to contain sixty-four per cent antipyrin, twenty-four per cent chlorine, and twelve per cent iron. A reddish, non-hygroscopic powder, used internally for anæmia, migraine, and neuralgia, externally in epistaxis, gonorrhœa, and local hemorrhages. Internal dose, gr. v.–xv. in solution. For external use, one to twenty per cent solution.

Guaiperol (Guaiacolate of Piperidine).—An English preparation recommended for phthisis. It is claimed to possess the antiseptic properties of guaiacol with the vascular tonic action of piperidine. Dose, gr. x. twice daily in cachets.